T0140294

Springer

Berlin
Heidelberg
New York
Barcelona
Hong Kong
London
Milan
Paris
Tokyo

Claus Rautenstrauch
Ralph Seelmann-Eggebert
Klaus Turowski
Editors

Moving into Mass Customization

Information Systems
and Management Principles

With 13 Figures
and 66 Tables

 Springer

Professor Dr. Claus Rautenstrauch
University of Magdeburg
Institute of Technical Information Systems (ITI)
Faculty of Computer Science
Universitätsplatz 2
39106 Magdeburg, Germany

Dipl.-Ing. Ralph Seelmann-Eggebert
Fraunhofer Institute for Factory Operation and Automation
Information, Logistics and Automation Systems
Sandtorstrasse 22
39106 Magdeburg, Germany

Professor Dr. Klaus Turowski
University of Augsburg
Chair of Business Information Systems,
esp. Application Engineering
Universitätsstrasse 16
86135 Augsburg, Germany

ISBN 3-540-43611-1 Springer-Verlag Berlin Heidelberg New York

Library of Congress Cataloging-in-Publication Data applied for
Die Deutsche Bibliothek – CIP-Einheitsaufnahme
Moving into Mass Customization: Information Systems and Management Principles;
with 66 Tables / Claus Rautenstrauch ... ed. – Berlin; Heidelberg; New York; Barcelona;
Hong Kong; London; Milan; Paris; Tokyo: Springer, 2002
    ISBN 3-540-43611-1

Springer-Verlag Berlin Heidelberg New York
a member of BertelsmannSpringer Science + Business Media GmbH

http://www.springer.de
© Springer-Verlag Berlin Heidelberg 2002
Printed in Germany

Cover design: Erich Kirchner, Heidelberg

SPIN 10878277   42/2202-5 4 3 2 1 0 – Printed on acid-free paper

Foreword

Mass customization is, undoubtedly, an engaging challenge nowadays, requiring a major combination of efforts in all areas of business. From sophisticated marketing concepts, to organizational issues, including logistics and manufacturing strategies, every field of activity has to be linked together to achieve what may be described as the ultimate form of competition. The concept of mass customization was first discussed in the beginning of the 90's of the last century, but the big take off was definitely caused by the growing role of electronic business as well as the application of Internet technology. Today we can say Internet is a powerful tool to implement mass customization.

However, mass customization must not be treated as just an "add-on" to be applied to conventional business concepts and strategies. It means a deep change in all parts of an operational enterprise. Therefore business processes, marketing concepts, company organization structure, information systems, production strategies, logistics, production technology, etc. have to be reconsidered. Furthermore, all subsystems of the supply chain have to work together in a continuously integrated manner. And, last but not least, in mass customization the human factor acquires a growing role ("robots don't make suggestions", Robert Pine II).

This book mainly aims to focus the strategic and technical aspects of mass customization. It is based on the results of two workshops organized by Claus Rautenstrauch and Klaus Turowski. The first one was the workshop *Mass Customization Management (MCM)* hosted by the *International Conference on Intelligent Systems & Application (ISA'2000)* in *Wollongong, Australia*, held in December 2000. In this workshop, a wide discussion about manufacturing issues took place. The workshop *Information Systems for Mass Customization (ISMC)* was hosted by the *International Conference on Information Science Innovations ISI'2001* and was held in March 2001, at *The American University of Dubai, Dubai, U.A.E.* The editors of this book would like to thank the organizers and chairmen of the conferences, *J. Ryffel, F. Naghd,, F. Kurfess, H. Ogata, E. Szczerbicki, H. Tlanfield* and *M. Sebaaly* for their extensive support. Furthermore, we thank *Werner A. Müller* for the opportunity to publish this book at Springer and his team for the editorial support they provided.

At last, we would also wish to show our deep appreciation to all authors from nine different countries, for their high quality papers and personal involvement.

This book was only made possible with the intense support of *C. Kunert* and *V. Rollert*.

Claus Rautenstrauch

Ralph Seelmann-Eggebert

Klaus Turowski

Magdeburg and Munich, August 2001

Table of Contents

Part I: THE STRATEGY OF MASS CUSTOMIZATION

Part II: PREPARATION AND IMPLEMENTATION

Part III: A VALUE CHAIN

Part I

The Strategy of Mass Customization

On the Economics of Mass Customization

Knolmayer, G. F.

Summary: As many "management philosophies" and business strategies, mass customization is often presented as a panacea for solving business problems. However, from the viewpoint of economic theory, there are almost no concepts that strictly dominate the previous situation. This holds also for the concept of mass customization. The paper discusses the pros and cons of mass customization and develops a model to explain the factors that influence the economics of this concept. The classical Cournot model of determining the maximum profit in a monopolistic situation is enlarged to deal with situations in which mass customization is a possible action, providing additional revenues, but also resulting in additional costs. Formulas for determining the effects of mass customization are developed and the optimal solution is computed.

Keywords: Mass customization; Strategy; Decision model; Cournot model.

1 Introduction

Mass customization is a synthesis of two long-competing management systems: Mass production systems versus those that provide products and services individually. The mass customization paradigm is based on flexibility and quick responsiveness in developing, producing, marketing, and delivering products that can satisfy as wide a range of customers as possible without substantially increasing costs. An often expressed viewpoint is that in the information age mass customization is replacing the mass production concept of the industrial age [Fitch/Johnson/Jones/Scyoc (2000); Interactive Custom Clothes Company (2000)].

In connection with "management philosophies" or "buzzwords" of the IT industry, new concepts are often propagated without too much reflection. Thus, quite commonly it is recommended to achieve a (mathematically) extreme value like "lot size 1" or "zero inventories".

From an economic perspective there are usually contradicting goals that are achieved by a management policy to certain extents. Management is responsible for decisions on policies and actions, based on their pros and cons, (at least implicitly) evaluating their benefits and costs. This is also true for a strategy like mass customization.

This paper reviews the potential benefits and costs of mass customization as a basis for determining whether it is profitable in a given competitive setting to provide a customization offering for those clients that are interested in it. A simple decision model based on Cournot's model of profit maximization in monopolistic markets is developed to explain decisions on mass customization strategies. An example shows that the numerical values of the parameters decide whether mass customization appears as an attractive strategy.

2   Customer Satisfaction and Mass Customization

Customer focus and satisfaction are main management concerns. However, these goals can be achieved in many different ways that are intensively discussed in connection with Customer Relationship Management (CRM) [Wayland/Cole (1997); Brown (2000); Newell (2000)].

Mass customization is one of several means for achieving higher customer satisfaction, but it may not always be perceived well. Firstly, law, ethics, and/or professionalism may prohibit customization offerings: No pharmacy can allow a seriously sick person to specify the ingredients for mixing a medicine to treat this illness. Incompetent customization may be less relevant in areas like beauty care for which Reflect.com offers products that "don't exist until you help create them".

Second, one has to consider whether the customers are really interested in specifying the requirements, e.g., of the mixture of elements in the tires they are using on their cars, although some customization options seem to emerge even for such types of products [Piller (2000)].

In general, the client of a product has a utility function in which at least two elements are relevant: The value of the products and services he is considering to buy and the time needed to handle the buying transaction and to receive the product. This viewpoint is closely related to economic theory which visualizes the tradeoff between income and leisure by indifference curves. Many clients will by no means be interested in specifying product features if this task is complex and time-consuming; a customization strategy is worthless for these types of customers and products. Consumer appreciation of customization offerings depends also on the frequency of purchasing a certain type of product [Kahn (1998)].

Sometimes even additional costs occur for the client who is opting for customization, e.g., if a configuration software has to be bought by the customer. Thus, a main parameter of the decision about customization strategies is the percentage of customers interested in specifying the features of a certain product. The "variety voice of the customer" has been proposed as a measure of the importance of a component's variety to the market [Martin/Hausman/Ishii (1998), p. 119].

Another determinant for mass customization is the price the customer accepts to pay for product differentiation. Homines oeconomici would often be better off, from a "rational" point of view, if they would accept an "umbrella variant" offering many options even if not all of them are on their list of beneficial features. Firstly, the customers may learn to esteem some options they did not cherish at the time of buying. Second, specialized plant designs and operations may result in (much) lower production costs due to (large) economies of scale.

However, not only the features of the product but also the chance to differentiate the product chosen from variants owned by other people may constitute a value for a customer for which he is willing to pay a premium. In many cases, customization of consumer products may simply be regarded as a gag, e.g., to make a present appear more personal. Lifestyle products seem to be the most suitable product category for this type of customization.

Thus, in connection with mass customization the following questions are relevant [Pine (1993), p. 167]:

- What about it is inherently personal so that it can differ for each individual?

- What do customers do differently with my product/service?

- How do customers differ in where they buy, receive, and use it?

- How do customers differ in when they buy, receive, and use it?

- How do customers differ in why they buy, receive and use it?

- How do customers differ in how they buy and use it?

3   Mass Customization, Technology, and Costs

Often the progress in Information Technology (IT) is regarded as a central means for making mass customization approaches feasible. In many cases, the Internet allows a direct communication with the user and he may be able to specify product features via the web [Baker/Van Dyke Parunak/Erol (1999)]. However, many integrity constraints exist for technically feasible product configurations and some type of artificial intelligence solutions may be necessary to guarantee that specifications of a non-expert customer are technically feasible. There may also be some bundling of features offered at the web to reduce the complexity of the specification process. The customer may be annoyed if only a very small amount is credited for abstaining from a certain component of the bundle.

Customization is rather easy if the service can be delivered digitally, e.g., via the web [cf. Schackmann/Link (2001)]: Personalizable web sites are an example for environments where the costs of customization are comparatively low and the benefits for an information-overloaded user are high. Personal editions of newspapers, journals, and even books have become quite common.

However, ease of communication is by no means a sufficient argument for introducing mass customization. Extreme flexibility of production facilities is much more important. In connection with Just-in-Time-philosophies, much effort has been devoted to cut setup times [Monden (1983), pp. 75]. Such industrial engineering projects have been very successful in some settings and without recognizable effects under other technical-organizational circumstances. Very high flexibility in technical means, procedures, organization, and convenient communication capabilities between customer and manufacturer may be necessary to make mass customization feasible and successful. The complexity of handling customized orders may be quite troublesome and involve high cost [NN (1998)]. In general, the costs of automation and of complex logistical systems have to be balanced against the advantages resulting from small lot sizes in the markets and in cutting throughput times and inventories [Bigler/Hopkins/Hopkins (1997)]. For the American automotive industry high risks of changing to a build-to-order strategy, a variant of mass customization, are anticipated; such a move would require "enormous changes in design, sourcing, manufacturing, marketing, and distribution" [Agrawal/Kumaresh/Mercer (2001)].

It has sometimes been argued that variants and mass customized products differ in the aspect whether the order of the client is already known or not. However, if postponement strategies [Lee/Tang 1997; Lee (1998)] are applied in producing variants, this distinction does not hold. Whereas some disadvantages of anonymous variant production (like obsolescence of stock and high return rates) may be avoided by mass customization, other disadvantages like high investments in factory automation and logistical complexity exist in both types.

Mass customization leads at least in the final operations of a multistage production process to extremely small lot sizes and may enlarge the complexity of logistics management considerably. Production management is faced with an ever increasing complexity in many areas and complexity management has become an important topic on the agenda of managers, consultants, and management scientists [Meijer (1998)]. One approach is trying to model the complexity of real-world systems in complex information and decision support systems. Another option is to start with trying to reduce the complexity of real-world systems [Burbidge (1962), p. 155]; these two options are visualized in Figure 1. Management of complexity is increasingly perceived as a strategic management task [Luczak/Fricker (1997)] and simplicity as a new competitive advantage [Jensen (2000)].

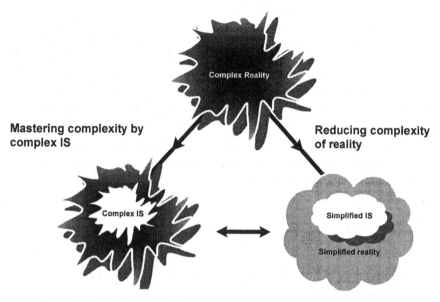

Figure 1: Strategies for handling complexity [Knolmayer/Mertens/Zeier (2001)]

At least one consulting firm employs the notion "complexity management" in its name and offers a "Complexity Manager" tool [GPS Prof. Schuh Komplexitäts-management]; modules of this package concentrate on visualizing the variant tree and on business games that focus on variant reduction. The management of product variety is regarded as an important subtask of complexity management. The Variety Effectiveness Program (VEP) is another method for helping companies decrease the complexity of variety [Galsworth (1994)].

Several studies have tried to determine the cost effects of offering a high product variety [for an overview cf. MacDuffie/Sethuraman/Fisher (1996)]. Baumhardt (1986) compared the total work content per average car, based on the actual vehicle mix, with the alternative of producing all cars with all options as standard, i.e., with the umbrella variant. If the standard work content for producing a customized car with all options is set to 100%, the work content is 89% for the actual product mix, and 90% for building all cars as the umbrella variant. Thus, the complexity cost in the existing plant is 10%. However, if the whole plant organization could be adjusted to the umbrella strategy, the work content is assumed to be only 69% and the complexity effect is estimated as 31%.

4 Models for Evaluating Mass Customization

The pros and cons of changing the number of product variants and lot sizes have been widely discussed in industrial engineering communities. On this basis, we compile in Table 1 main arguments in favor and against mass customization. A manager will have to (at least implicitly) put weights to the different items mentioned in Table 1 to decide whether mass customization is attractive in a certain business setting. In general, the conclusion is less straightforward as many authors and consultants present it.

One quantitative approach to discuss the consequences of variety are scores computed by methods like "Design for Product Variation" or "Design by Variety" [Martin/Ishii (1996); Martin/Ishii (1997)]. Three features are used in this analysis:

- Commonality (M1)
- Differentiation point (M2)
- Set-up cost (M3).

They are measured at a {0, 1} scale in such a way that lower values signify less attractive attribute values.

Thus, the disadvantages of variety DV may be expressed by

$$DV = 1 - M_1 * M_2 * M_3$$

Table 1: Pros and cons of mass customization [cf. Knolmayer (1999)]

PRO MASS CUSTOMIZATION	CONTRA MASS CUSTOMIZATION
• Diversification strategy.	• Cost leadership strategy.
• Unique selling position allows precise fulfillment of uncovered customer needs and wishes, leads to high acquisition potential, and allows a quasi-monopolistic price policy.	• Avoid additional costs in client-counseling, selling, design, documentation, disposition, administration, control, production, logistics, service, and recycling.
• Individualized and long-term customer relationship; low marketing expenditures, low inventory, low return rate; reduced number of products that have to be sold at large discounts to clear inventories.	• Avoid problems in defining and configuring product variants.

• No need to separate manufacturing and retailing; vertical integration.	• Avoid complex coordination between sales force, design, production, and logistics.
• Delivered product includes only features that are regarded as relevant by the client.	• Customer needs may evolve over time and the emerging needs may be satisfied by "umbrella-variants".
• Low complexity of using the product because it fits directly to client's needs.	• Difficulties in providing customized documentations and user manuals.
• Knowledgeable workers needed; motivation of employees by job enlargement.	• Avoid problems in after-sales service and in providing adequate spare parts.
• Avoid high ad-hoc costs if customization is exceptionally inevitable for a mass manufacturer.	• Avoid logistical problems in controlling the material flow, uneven balance of production facilities, and irregular material flow.

The result of this computation is confronted with the importance of variety from the viewpoints of the customers in a portfolio diagram. In another, additive model it is suggested to estimate the parameters of

$$DV = \alpha + \beta * M_1 + \gamma * M_2 + \delta * M_3$$

by regression analysis [Martin/Ishii (1996)].

The following elementary microeconomic analysis of mass customization is based on the variables, parameters, and indices defined in Table 2.

Table 2: Symbols and numerical values used in the example.

Symbol	Definition	Determined by	Numerical value
a > 0	Parameter in price/demand function for standardized product		1000
b > 0	Parameter in price/demand function for standardized product		1

c > a	Parameter in price/demand function for customized product	(11)	1100
D	Parameter in price/demand function for customized product	(12)	1
F	Fixed annual cost without providing customization		50000
F' > F	Fixed annual cost with providing mass customization		80000
g	Quota of additional clients due to customized offerings		0.25
K	Direct costs of producing one standard unit (without providing customization)		200
k' > k	Direct costs of producing one customized unit		220
Q	Quota of clients preferring the standard product		0.8
z > 0	Factor for price increase allowed by customization		0.1

We assume that a monopolistic price/demand function

$$p = a - bx \qquad (1)$$

(with a>0, b>0) exists if a product is only available in a standardized version and no customization is offered.

This results in revenues

$$R = (a - bx) \, x \qquad (2)$$

and the profit becomes

$$PR = [(a - bx) \, x] - F - k \, x \qquad (3)$$

We assume that the goal of the firm is to maximize profits. In this case, the optimal solution can be determined as

$$x^* = (a - k) / (2b) \qquad (4)$$

The resulting price is

$$p^* = (a + k) / 2 \qquad (5)$$

and the maximum profit becomes

$$PR^* = [(a - k)^2 / (4b)] - F \qquad (6)$$

Let $0 < q < 1$ be the quota of clients preferring to buy the standard version s and $(1-q)$ the quota who are interested in customized solutions c.

We assume that there are no interdependencies between the two market segments. Thus, it is not possible to rise p because after offering customized products a smaller number of clients is demanding the standard variant and it is also not necessary to lower the price because the total number of clients may rise due to the customization option.

The price/demand function for the buyers of the standardized version is

$$p_s = a - b \, x_s / q \qquad (7)$$

because we assume that they do not change their buying behavior when some other buyers switch to customized solutions. Thus, the optimal price (5) remains unchanged but a smaller number of products is sold at this price:

$$x_s^* = (a - k) \, q / (2b) \qquad (8)$$

The resulting gross margin is

$$GM_s^* = [q \, (a - k)^2 / (4b)] \qquad (9)$$

The price/demand function of the clients favoring customization may be analogously expressed as

$$p_c = c - d \, x_c \qquad (10)$$

We assume that the clients who are interested in customized products are willing to pay a premium factor of $z > 0$, resulting in

$$c = a \, (1 + z) \qquad (11)$$

We further assume that at the previously optimal price (5) the company can gain an additional quota of $0 < g < 1$ customers due to customization. Thus we know two points of the linear function and can determine its descent by

$$d = [(2c - a - k) \, b] / [(1 + g) \, (a - k)] \qquad (12)$$

In computing the optimal solution we must consider also the demand of customers preferring the standardized version.

To allow mass customization, additional investments are usually necessary. We assume that either the whole factory or no department of it will be prepared for mass customization. Let the annual fixed costs associated with making the organization ready for mass customization be $F' > F$ and the variable costs for providing a unit of the customized product k'>k; the additional costs implied by mass customization are emphasized by Graff (1994). The optimal solution if all clients would prefer customization (i.e., if q=0) becomes

$$x_c^* = [(c - k') (1 + g) (a - k)] / [2b (2c - a - k)] \tag{13}$$

$$p_c^* = c - d \, x_c^* = (c + k') / 2 \tag{14}$$

Because we assume that no interdependencies between the two markets exist, we obtain

$$x_c^* = [(c - k') (1 + g) (a - k) (1 - q)] / [2b (2c - a - k)] \tag{15}$$

$$R_c^* = [(c^2 - k'^2) (1 + g) (a - k) (1 - q)] / [4b (2c - a - k)] \tag{16}$$

The maximum value of the objective function is based on the profits that can be obtained with and without mass customization. Thus one has to compare (6) to

$$PR_{sc}^* = [q (a - k)^2 / (4b)] + [(c^2 - k'^2) (1 + g) (a - k) (1 - q)) / (4b (2c - a - k)] -$$

$$F' - [(k' (c - k') (1 + g) (a - k) (1 - q)) / 2b (2c - a - k))] \tag{17}$$

The model is exemplified by using the data given in Table 2.

If only a standardized product is offered, we obtain the following results:

$$p = 1000 - x$$

$$x^* = 400$$

$$p^* = 600$$

$$PR^* = 110{,}000$$

This result is visualized in Figure 2.

For customization the price/demand function is derived as

$$p_c = 1100 - x_c$$

Figure 3 shows the price/demand functions without customization (lower curve) and under the assumption that all clients would prefer customized to standard products.

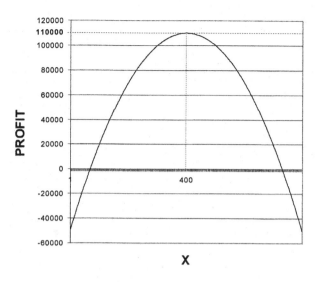

Figure 2: Profit as a function of the number of units sold

If all clients would opt for customization the solution

$x_c^* = 440$

$p_c^* = 660$

could be obtained by (13) and (14). However, if only 20% of the customers show interest in customized products, we obtain by (15) and (16)

$x_c^* = 88$

$R_c^* = 58,080$

By (17) we compute the maximal profit obtainable with partial mass customization as

$PR_{sc}^* = 86,720 < PR^* = 110,000$

Thus, in this example the strategy of mass customization appears as unattractive.

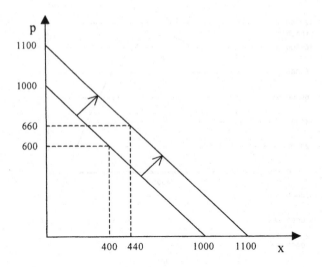

Figure 3: Price/demand functions without and with customization

Of course, for some parameter values customization is profitable and for other ones it does not pay. For instance, if one would assume q=0.1 and F'=60,000, the result would be

$$PR_{sc}^* = 120,240 > PR^* = 110,000$$

In this case the large majority of clients that prefer customizing should be provided with this option by a profit-maximizing firm.

Main determinants of selecting a mass-customization strategy are

- the reaction of the market to the customized offerings,
- the investment necessary to realize a mass-customized production,
- the effect of customization on direct costs, and
- the severity of inefficiencies resulting from wrong forecasts in a mass make-to-stock production.

5 Summary and Outlook

The paper shows that mass customization has advantages and disadvantages and that it is necessary to provide criteria and decision support models for determining which product groups should be prepared for customization and for which groups standardized products should be supplied. Arguments pro and contra customization are presented in a systematic manner and a microeconomic model is developed showing the determinants of selecting a mass customization strategy.

A more detailed analysis would have to distinguish several product groups that may be either offered as customized products or not. Fixed costs of making parts of the factory and the organization ready for mass customization may be shared by different product groups. A mathematical programming model could be formulated which suggests allocations of investments for mass customization to different shops of the plant with the goal of determining the optimal extent of customization.

References

Agarwal, M., Kumaresh, T.V., Mercer, G.A. (2001): The false promise of mass customization. McKinsey Quarterly, Nr. 3, 62-71
http://www.mckinseyquarterly.com/article_page.asp?tk=317344:1089:2&ar=1089&L2=2&L3=38

Baker, A.D., Van Dyke Parunak, H., Erol, K. (1999): Agents and the Internet: Infrastructure for mass customization. IEEE Internet Computing 3/5, 62-69

Baumhardt, J.J. (1986): The Effect of Complexity on Product Costs and Profits. In: Wildemann, H. (Ed.): Strategische Investitionsplanung für neue Technologien in der Produktion, Volume 2. gfmt, Passau, 779-810

Bigler, L.R., Hopkins, S.A., Hopkins, W.E. (1997): Mass customization in high tech/highly complex product environments: Reconciling the high cost-small lot size dilemma. In: 1997 Proceedings. Decision Sciences Institute, 28th Annual Meeting, Volume 3. Decision Science Institute, Atlanta, 1150-1152

Brown, S.A. (2000): Customer Relationship Management - A Strategic View in the World of e-Business. Wiley Canada, Toronto et al.

Burbidge, J.L. (1962): The Principles of Production Control. Macdonalds and Evans, Estover

Fitch, K., Johnson, P., Jones, J. ,Scyoc,J. (2000): MassCustomization
http://www.campbell.berry.edu/faculty/jgrout/bus453/masscustomization/mass.html [2000-03-23]

Galsworth, G.D. (1994): Smart, simple design: using variety effectiveness to reduce total cost and maximize customer selection. Wight, Essex Junction

16

GPS Prof. Schuh Komplexitätsmanagement GmbH: Willkommen auf
www.ComplexityManager.de_http://www.complexitymanager.de/

Graff, J. (1994): Mass Customization and Fractalization, A Review
http://users.iems.nwu.edu/~jamison//quartz/masscust.html

Interactive Custom Clothes Company (2000): MassCustomization
http://www.ic3d.com/2000/aboutus/mass_customization.html [2000-05-02]

Jensen, B. (2000): Simplicity - The New Competitive Advantage in a World of More, Better, Faster. Perseus, Cambridge

Kahn, B. (1998): Variety: From the Consumer's Perspective. In: Ho, T.-H./Tang, C.S. (Eds.): Product Variety Management, Research Advances. Kluwer, Boston et al., 19-37

Knolmayer, G. (1999): Kundenorientierung, Mass Customization und optimale Variantenvielfalt. In: Grünig, R., Pasquier, M. (Eds.): Strategisches Management und Marketing. Haupt, Bern et al., 67-91

Knolmayer, G., Mertens, P., Zeier, A. (2001): Supply Chain Management Based on SAP Systems. Springer, Berlin et al. (to appear)

Lee, H.L. (1998): Postponement for mass customization. In: Gattorna, J. (Ed.): Strategic Supply Chain Alignment. Gower, Brookfield, 77-91

Lee, H.L., Tang, C.S. (1997): Modelling the Costs and Benefits of Delayed Order Differentiation. Management Science 43 / 1, 40-53

Luczak, H., Fricker, A. (1997): Komplexitätsmanagement - ein Mittel der strategischen Unternehmensgestaltung. In: Schuh, G., Wiendahl, H.P. (Eds.): Komplexität und Agilität. Springer, Berlin et al., 309-323

MacDuffie, J.P., Sethuraman, K., Fisher, M.L. (1996): Product Variety and Manufacturing Performance: Evidence from the International Automotive Assembly Plant Study. Management Science 42 / 3, 350-369

Martin, M., Hausman, W., Ishii, K. (1998): Design for Variety. In: Ho, T.-H./Tang, C.S. (Eds.): Product Variety Management, Research Advances. Kluwer, Boston et al., 103-122

Martin, M.V., Ishii, K. (1996): Design for Variety: A Methodology for Understanding the Costs of Product Proliferation. In: Proceedings of The 1996 ASME Design Engineering Conferences and Computers in Engineering Conference, Irvine
http://mml.stanford.edu/Research/Papers/1996/1996.ASME.DTM.Martin/1996.ASME.
DTM.Martin.pdf

Martin, M.V., Ishii, K. (1997): Design for Variety: Development of Complexity Indices and Design Charts. In: Proceedings of DETC'97, 1997 ASME Design Engineering Technical Conferences, Sacramento
http://www-mml.stanford.edu/Research/Papers/1997/1997.ASME.DFM.Martin/1997.
ASME.DFM.Martin.pdf

Meijer, B.R. (1998): To manage or not to manage complexity. In: IEMC '98 Proceedings. International Conference on Engineering and Technology Management. Pioneering New Technologies: Management Issues and Challenges in the Third Millennium. IEEE, New York, 279-284

Monden, Y. (1983): Toyota Production System, Practical Approach to Production Management. Institute of Industrial Engineers, Atlanta

Newell, F. (2000): loyalty.com - Customer Relationship Management in the New Era of Internet Marketing. McGraw-Hill, New York et al.

NN (1998): Mass Customization Technology
http://iesu5.ust.hk/research/mctcrc/mct.html [1998-09-15]

Piller, F. (2000): Modularisierung in der Reifenindustrie - die „reverse economy". Mass Customization News 3 / 6
http://www.mass-customization.de/news/news00_06.htm#1

Pine, B. J. (1993): Mass Customization. Harvard Business School Press, Boston

Schackmann, J., Link, H. (2001): Mass Customization of Digital Products in Electronic Commerce. In: Seebaly, M.F. (Ed.): Proceedings of the International NAISO Congress in Information Science Innovations (ISI'2001). ICSC Academic Press, Slierecht et al., 144-150

Wayland, R.E., Cole, P.M. (1997): Customer Connections - New Strategies for Growth. Harvard Business School Press, Boston

A Procedure for Building Product Models

Hvam, L., Riis, J., Malis, M. & Hansen, B.

Summary: This article presents a procedure for building product models to support the specification processes dealing with sales, design of product variants and production preparation. The procedure includes, as the first phase, an analysis and redesign of the business processes, which are to be supported with product models. The next phase includes an analysis of the product assortment, and the set up of a so-called product master. Finally the product model is designed and implemented using object oriented modelling. The procedure is developed in order to ensure that the product models constructed are fit for the business processes they support, and properly structured and documented, in order to facilitate that the systems can be maintained continually and further developed. The research has been carried out at the Centre for Industrialisation of Engineering, Department of Manufacturing Engineering, Technical University of Denmark.

1. Introduction

A product model supports the activities of specifying products in sales, design and methods engineering – the specification process. The specification process denotes the part of the engineering system where the specifications for the customised product variants are created as illustrated in Figure 1.

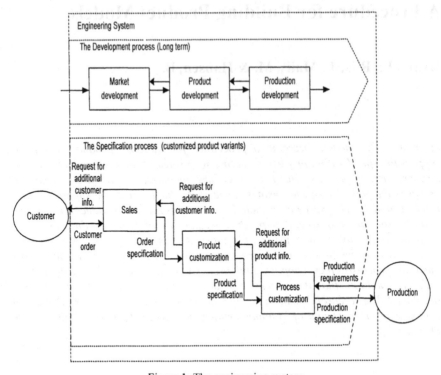

Figure 1: The engineering system

The activities of specifying products can be supported by implementing product models containing knowledge and information on products and models containing knowledge on related systems, like e.g. production or assembly. Knowledge integrated product- and product related models are defined as:

A knowledge base which contains part of or all of the knowledge and information associated with the product in different phases of the product's life cycle, e.g. sales, design, production, assembly, service and reuse.

Product related models contain knowledge and information on the systems related to the product's life cycle, while the product model contains knowledge and information on the product's structure and functional properties [Krause, 1988].

The use of product and product related models is increasing. There are today a number of examples of product and product related models which, for instance, are used to support sales, design of product variants and production preparation. Experience from a considerable number of Danish companies shows that often these models are constructed without the use of a strict procedure or modelling techniques. The result is often that the systems are unstructured and undocumented and they are therefore difficult or impossible to maintain or further develop. Thus there is a need to develop a procedure and associated modelling techniques, which can ensure that the constructed product and product related models are properly structured and documented, so that the systems can be maintained continually and developed further.

Another experience is that the product and product related models are not always designed to fit the business processes, which they are meant to support. Finally an important precondition for building product models is that the products are designed and structured in a way, which makes it possible to define a general master of the product, from which the customer specific products can be derived.

In order to cope with these challenges, a procedure for building product models should include: An analysis and redesign of the specification processes in focus, an analysis and eventually redesign/ restructuring of the products to be modelled, and finally, a structured "language" - or modelling technique - which makes it possible to document the product and product related models in a structured way.

2   A Procedure for Building Poduct Models

Figure 2 shows a procedure for building product and product related models. The procedure contains 7 phases. The starting point for the work is an analysis and redesign of the business processes, which will be affected by the product and product related models (phase 1). In phase 2 the products are analysed and described in a so-called product master. Phase 3 includes the final design of the product and product related models using the object oriented modelling techniques. Phases 4 to 7 deal with design, programming, implementation and maintenance of the product models. Phases 3 to 7 follow by and large the general object oriented project life cycle.

There may be some overlap and iterations between the individual phases.

Phase	Description
1	**Process Analysis** Analysis of the existing specification process (AS-IS), statement of the functional requirements to the process. Design of the future specification process (TO BE). Overall definition of the product – and product related models to support the process. **Tools:** IDEF0, flow charts, Activity Chain, Model, key numbers, problem matrix, SWOT, list of functional describing characteristics and gap analysis.
2	**Product Analysis** Analysing products and eventually life cycle systems. Redesigning/ restructuring of products. Structuring and formalising knowledge about the products and related life cycle systems in a product master. **Tools:** List of features and product master.
3	**Object Oriented Analysis (OOA)** Creation of object classes and structures. Description of object classes on CRC-cards. Definition of user interface. Other requirements to the IT solution. **Tools:** Use cases, screen layouts, class diagrams and CRC-cards.
4	**Object Oriented Design** Defining and further developing the OOA-model for a specific programming tool.
5	**Programming** Programming the system. Own development or use of standard software.
6	**Implementation** Implementation of the product- and product related models in the organisation. Traning users of the system, and further training of the people responsible for maintaining the product- and product related models
7	**Maintenance** Maintenance and further development of the product and product related models.

Figure 2: A procedure for building product models. The contents of the phases are described in the following.

2. 1   Phase 1 – Process Analysis

Initially an AS-IS description of the current processes is made. This description should expose the structure between activities, people, IT systems, shifts of responsibility etc. Key figures for characterising quality, resource consumption and throughput times may support the description. Analysis tools to be used may be IDEF0 [ICAM, 1981], The Activity Chain Model [Barfod, 1997], or different kinds of flow charts.

In the other part of the process analysis the future requirements set by the surroundings are analysed and determined making it possible to evaluate the gap between the current process performance and the required performance. To support the requirement analysis a list consisting of functional characteristics is used. These are among others:

- Kinds of input and output specifications
- Throughput time
- Resource consumption
- Quality of specifications
- Insight into consequences
- Flexibility of the process
- Frequency of similar specification activities
- Accessibility of knowledge
- etc.

The functional characteristics are further outlined in [Hvam and Hansen, 1999 and Hvam et al 2000]. Based upon the functional requirements to the specification process and the characteristics of the existing specification process, a gap analysis is made in order to identify the major gap between the current performance of the business process and the requirements to the process. This leads to identify the potential improvements to realise using product- and product related models.

Based on the process analysis a concept for a future ideal business process is designed. This ideal concept is made in order to be more creative and not so restricted by "historic" procedures in the company, - similar to the BPR approach [Hammer, 1990]. With the ideal concept in mind a more realistic process design (TO-BE) is made. This TO-BE description will consist of structural elements such as:

- Process structure
- Humans

- Organisation

- IT-systems

In relation to the definition of the future specification process, the product- and product related models to support the specification process are defined in overall. Finally the purpose, view and context of the models to be build are defined.

2.2 Phase 2 – Product Analysis

In this phase the product to be modelled is analysed in order to gain an overview of the product families, and their structure. The analysis covers the product's function and structure, the properties of the product, the variations of the product and the related systems in the product's life cycle. Figure 3 shows a general architecture for describing products including the above mentioned views.

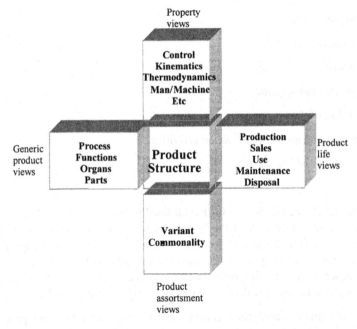

Figure 3: An architecture for describing products [Hubka, 1988], [Mortensen, 2000]

Normally, product and product related models only contain a minor part of the proposed views in the architecture. The specific views to include in the models are defined based on the overall content of the product and product related models outlined in phase 1.

Before the object-oriented analysis (OOA) is carried out, an overview over the product assortment and other views is set up by use of a so-called product master. Figure 4 below shows a part of a product master for a bookcase.

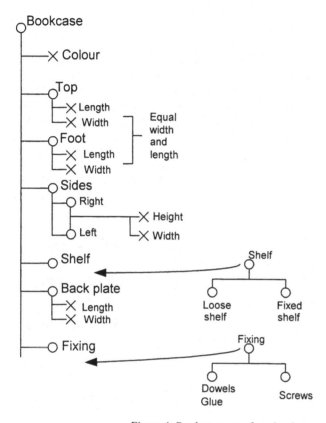

Figure 4: Product master for a bookcase

The product master is build up from part-of structures and kind-of structures. A cross indicates a part in the bookcase, while a circle indicates a characteristic (attribute in the object oriented modelling language). Relations/ constraints between the parts are indicated by a line between the two parts, and the relation is described.

2.3   Phase 3 – Object Oriented Analysis (OOA)

OOA is a method used for analysing a given problem domain and the field of application in which the IT system will be used. The purpose of the OOA is to analyse the problem domain and the field of application in such a way that relevant knowledge can be modelled in an IT system. The problem domain is the part of reality outside the system that needs to be administrated, surveyed or controlled. The field of application is the organisation (person, department) that is going to use the system to administer, survey or control the problem domain.

2.3.1 Modelling the problem domain:

The OOA model can be made through the activities described in figure 5, which describes the OOA as consisting of five phases or layers. These layers can be seen as different viewpoints, which together make up the OOA model. Normally the five activities are carried out through a top down approach, but there are no restrictions in that sense. Typically the OOA model will be the result of a number of iterations of the analysis process.

The *subject layer* contains a sub-division of the complete domain which is to be modelled in different subject areas. In relation to the use of product models, a subject area can for example be a product model or a factory model [Krause, 1988].

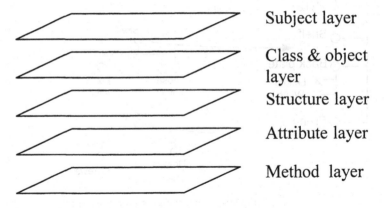

Subject layer

Class & object layer

Structure layer

Attribute layer

Method layer

Figure 5: The five layers of OOA modelling [Coad, 1991]

The *class and object layer* contains a list of the classes and objects which have been identified in the individual subject areas.

The *structure layer* contains the relationships between the objects, i. e. a specification of generalisation and aggregation.

The *attribute layer* contains a specification of the information associated with the individual objects, i. e what the objects know about themselves.

The *method layer* contains a description of the individual objects methods (procedures), i. e. what the objects can perform.

Classes and structures are identified based on the product master from phase 2. The static structure is mirrored in the layers of theme, classes and objects, structure and attributes, while the more dynamic aspects in the model mainly are related to the method layer. The result of the OOA can be illustrated in a class diagram [Booch et al, 1999] and on CRC cards [Hvam and Riis, 1999]. The notation used in the class diagram is illustrated in figure 6, which shows a class and four different structures. The notation is part of the Unified Modelling Language (UML), which has been chosen since it is the preferred standard world wide and is used in many development tools.

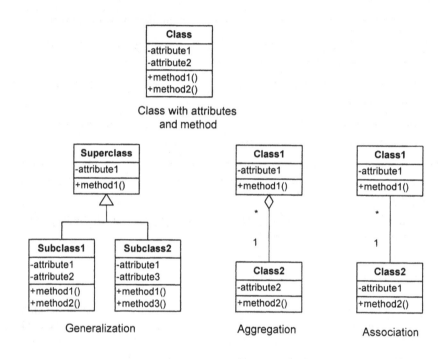

Figure 6: Notation for Class Diagram (UML)

2.3.2   Modelling the field of application

The second part of the OOA consists of an analysis of the field of application. Here the interaction between man and machine is analysed in order to determine the functionality of the system, the user interface, software integration to other IT-systems etc. Other elements that need to be determined are also requirements to response time, flexibility and so on. The result of this is a description of the user interface and a requirement specification for the product and product related models.

2.4   Phase 4 – Object Oriented Design

When a system is being built up, the perspective changes from being domain oriented (what and which task?) to be implementation oriented (how?). According to [Coad, 1991] four perspectives are used during development of the OOD model:

- User interface, which determines the user's communication with the system.

- Problem domain, in which the OOA model is corrected in accordance with design-specific criteria.

- Data management, where the structure of the stored data and methods for control of data are modelled.

- Task management, which is used in cases where the system has to perform several tasks simultaneously (multitasking).

Object-oriented design contributes, like the other phases of the object-oriented project life cycle, to a structured procedure, and thus makes it possible to control the entire project more closely.

If a standard configuration tool is used (Baan Configurator, Oracle SellingPoint, etc.) most of the design parameters are frozen.

2.5   Phase 5 – Programming

The selection between a standard software system or own development depends on the company's resources such as economy, programming skills, Current IT systems etc. If the company decides to develop its own system, object oriented programming languages such as C++, Java, or Smalltalk can be used.

In the last few years a large amount of work has been done to create various standard configuration solutions. The major suppliers of ERP and Front Office systems are now joining the market. The list below illustrates the most famous actors at the market:

Front Office/ERP:	Front Office:
Baan (now Invensys CRM), Oracle, Cincom, Sap, Intentia, i2 Technologies, J. D. Edwards, Peoplesoft etc.	Siebel, Trilogy, Calico, Firepond, Selectia, BT Squared, Clarify, SalesLogix etc.

A standard system makes the domain experts more independent of software programmers since the actual programming is relatively simple and can be done without extensive programming skills. However the integration of a standard system to other systems would normally necessitate the work from programmers.

Use of a standard system could give advantages such as: quick installation, easier development and maintenance, supplier support, safe IT costs, higher system- and education quality and possibility of a test period before purchasing. On the other hand a standard system can be quite expensive and may lead to some disadvantages such as supplier dependence, integration/fitting difficulties, and changed work terms.

2.6   Phase 6 – Implementation

Implementation, user acceptance, maintenance and follow-up are very critical factors. The system must stand its trial here. User acceptance is completely crucial if the system is to be a success; if the users are not satisfied, the system will not survive long. One way of getting the users' acceptance of the system is to involve the users already in the analysis phase. This can be done by developing an early prototype of the system, which the users can comment on. Also training and current information of the users will facilitate the users' acceptance of the system.

2.7   Phase 7 – Maintenance

Product and product related models can be viewed as "living organisms." The models will soon lose their value if they are not further developed and maintained. The object oriented structure and documentation (class diagrams, CRC cards etc.) of the product and product related models make it considerably easier to maintain and develop the models further.

Application of product modelling introduces a new way of doing business. New tasks are introduced in the organisation. E.g. a salesman will have to use the product and product related models in order to configure a product, and a product designer will have to build up and maintain the information and rules describing the products. This calls for commitment and ongoing motivation from the top management. Besides this both users and model managers need education and training in using and maintaining the product and product related models.

3 Empirical Study

The proposed procedure has been tested at a Danish wind turbine manufacturing company, NEG Micon A/S, from August 1999 to March 2000 [Mertz and Vølund, 2000]. The aim of the project at NEG-Micon was to construct a product model (sales configurator) for the sales process. The work is described in the following.

3. 1 Case Study Phase 1- Process Analysis

A model of the sales- and specification process is shown on figure 7. It illustrates the main activities from the customer request to the final documentation.

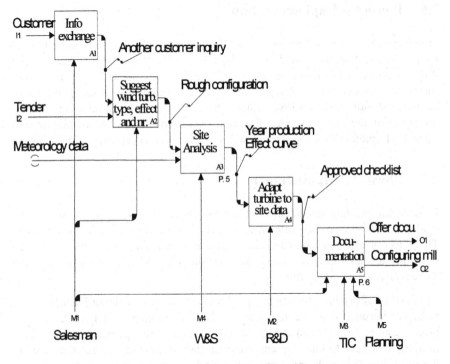

Figure 7: IDEF0 diagram of the sales and specification process

The analysis showed several weaknesses. It was for example found that approx. 55% of the specifications lacked quality. Based on the AS-IS description and the functional requirements gaps have been found in several areas:

- Quality of the specifications
- Resources

- Knowledge division

- Lead-time

Furthermore the process should be able to:

- Handle a larger number of variants

- Handle continuous product changes in a better way

- Include approvals and regulatory requirements

Based on the above analysis of the requirements to the business process a future business process was set up. Based on this a simple diagram of an ideal configuration system was sketched and discussed with the representatives of the company. Three limitations (functionality, price of the system and time of development) led to the adjusted model as illustrated on figure 8. Business Processes, which were difficult to support with product and product related models were sorted out.

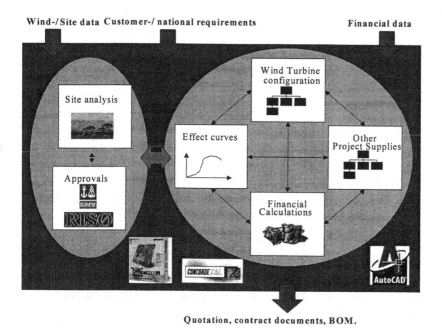

Figure 8: Adjusted Product Model

As a description tool for defining the interaction between the individual parts of the product and product related models and the people working in the specification process a Use Case has been made. The Use Case (figure 9) shows the interactions between the employees at NEG-Micon involved in the specification process and the proposed product and product related models.

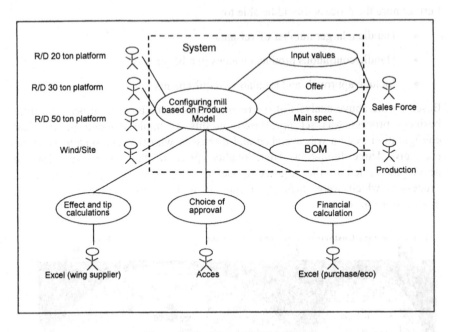

Figure 9: Use Case at NEG Micon

Based on the above listed analysis the purpose, view and context of the product and product related models have been defined as:

Purpose: The purpose is to set up a system, which is able to specify bills of materials at an overall level for 80% of the windmills sold.

View: The view is primarily the view of the sales personnel, secondary the view of the product designers who will be responsible for maintaining the rules in the system.

Context: The context of the models is specification of bills of materials at an overall level for windmills based on the 20 tons platform.

Standard software (Baan Configuration version 98.2) was chosen for the programming.

3. 2   Case Study Phase 2 – Product Analysis

Figure 10 shows a part of the windmill (Nacelle and rotor).

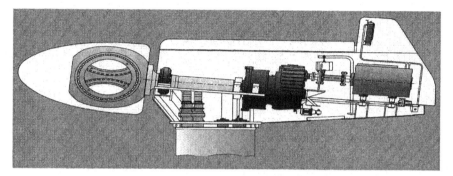

Figure 10: Nacelle and Rotor

In order to get an overview of the windmill the product was analysed and a product master was built up. The overall content and the details to include in the product master is decided based on the analysis of the business processes in phase 1. Figure 11 illustrates a part of the product master for windmills, where different parts of the wind turbine are described.

34

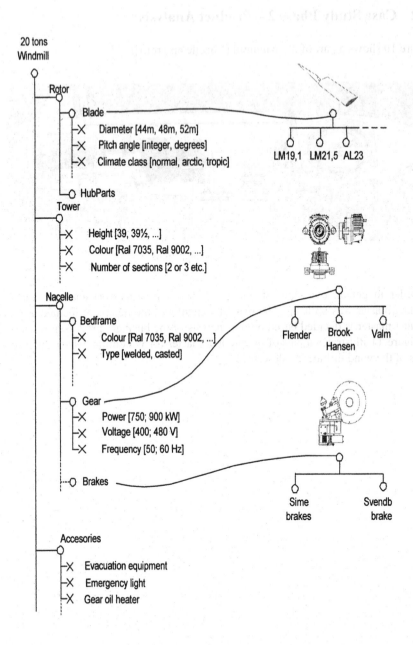

Figure 11: A part of the product master for a windmill

3. 3   Case Study Phase 3 – Object Oriented Analysis model

The OOA model is divided into five themes (generating documents, project information, configuration of the wind turbine, wind data and calculations). Each theme has classes connected. They are divided into attributes and methods. Figure 12 shows the contents of theme 3: Configuration of the wind turbine.

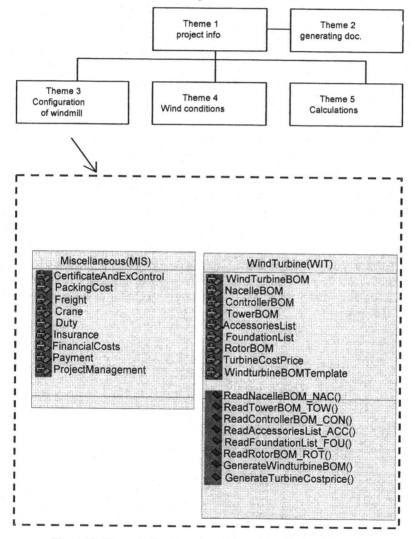

Figure 12: Theme 3: Configuration of the turbine. Rational Rose.

3. 4   Case Study Phase 4 – Object Oriented Design

In this case a standard configuration system (Baan Configurator) was used. As with most standard systems, the design is frozen by the supplier.

3. 5   Case Study Phase 5 – Programming

When programming a standard system the rules for programming are set by the supplier. The Baan Configurator is logic and constraint based. The product attributes and constraints are programmed based on the attributes and methods listed on the CRC-cards from the OOA-model.

The rules are constructed with logical operators (AND, OR, NOT, TRUE, FALSE). They can be divided into three categories: boolic, arithmetic and warnings. The source of the rule indicated by a number. As an example a rule concerning the rotor (with id number a1678900) is showed below:

Rotor_BOM[a1678900]Rotor_diameter[a44]AND
Windclass[IEC1_plus]ANDOneOf(Climate_class[Normal],
Climate_class[Tropical]...

Figure 13 shows an example of the system's user interface. Choices made by the user are marked by a check mark. If the selection is against the rules, then the configurator will give an automatic warning and a note that indicates where the selections must be corrected/changed. The user is able to start the configuration at a random place and it is possible to optimise according to different resources (price, output etc.).

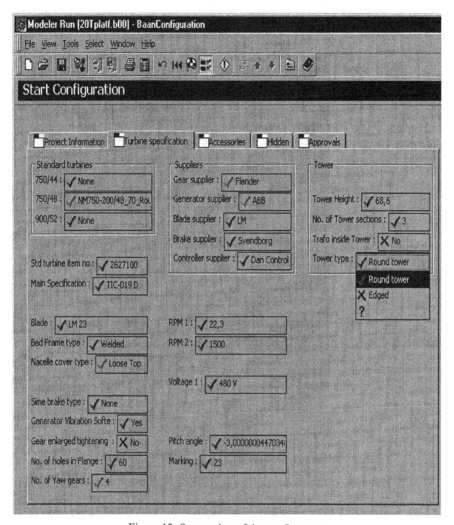

Figure 13: Screen shot of the configurator

3. 6   Case Study Phase 6 – Implementation

The final implementation of the product model is yet to be done. The prototype described in phase 6 shows a huge potential for implementing the product model. Several considerations must be done before a full-scale implementation is possible. A Cost-Benefit analysis of a full-scale implementation has been made. It is based on quantitative advantages, qualitative advantages, expenses for IT and internal activities. It is found that the payback period will be less than one year.

Based on the prototype described in this paper, the company has decided to continue the work and implement a product configuration system.

3.7 Phase 7 - Maintenance

The product master and the object oriented model provide a structure in the software and serve as documentation of the system.

4 Conclusion

The proposed procedure is based on well known and proved theory elements. The aim of the procedure is to serve as guidance for engineers working with product modelling. The procedure has been tested at several manufacturing companies in Denmark and abroad with positive results.

The proposed procedure includes several fields of expertise:

- Business process reengineering, and business strategy
- Product design and manufacturing technology
- Theory for structuring mechanical systems, and structuring product and product related models
- Object oriented modelling
- IT, Artificial intelligence and knowledge representation
- Organisational aspects of application of product modelling

The wide range of theory is included in the procedure in order to cope with the questions raised in the introduction of the paper. I.e. how to deal with the business processes affected by the models, how to analyse and structure products and how to implement the models in IT-systems.

Only few of us will claim to be an expert in all the fields mentioned. However engineers working with product modelling will need to obtain some insight in the fields listed. Therefore many engineers working with product modelling could benefit from qualifying themselves within themes of relevance for product modelling, where they have little insight.

Based on the proposed procedure a 4 days intensive course in product modelling has been developed. The aim of the course is to introduce the modelling techniques and analysis methods in the proposed procedure. The experiences from the course have been positive.

References

Andreasen, M.M. et al.: On structure and structuring Workshop – Fertigungsgerechtes Konstruieren, 1995.

Barfod, A. & Hvolby, H.: Ordrestyring - tidens indsatsområde, Industriens Forlag, 1997.

Booch, G., Rumbaugh, J. & Jacobson, I.: The Unified Modeling Language User Guide, Addison-Wesley, 1999.

Coad, P. & Yourdon, E.: Object-oriented analysis, second edition: Prentice Hall, New Jersey, 1991.

Hammer, M.: Reengineering Work: Don't Automate, Obliterate, Harvard Business Review, July-August, 1990.

Harlou, U.: A product family master plan as basis for product modeling and engineering design, 1999.

Hubka, V. & Eder, W.E.: Theory of Technical Systems, Springer-Verlag. Berlin, 1988.

Hvam, L.: Application of product modelling – seen from a work preparation viewpoint, Ph.D. thesis, Department of Industrial Management and Engineering, Technical University of Denmark, 1994.

Hvam, L. & Riis, J.: CRC Cards for Product Modelling, Department of Manufacturing Engineering, Technical University of Denmark, 1999.

Hvam, L., Hansen, B.L.: Strategic guidelines for application of product models; The 4th Annual International Conference on Industrial Engineering Theory, Applications and Practice, San Antonio, Texas, November 17-20 1999.

Hvam, L., Mortensen, N. H., Riis, J. and Hansen, B.L.: Produktmodellering – procesanalyse, produktanalyse, objektorienteret analyse, Department of Industrial Management and Engineering, Technical University of Denmark, 2000, 180 pages (In Danish).

ICAM project group: Integrated Computer-Aided Manufacturing (ICAM) Architecture part II, Vol. IV-Function Modelling Manual (IDEF-0); Soft Tech Inc, MA USA, Juni1981.

Johnson, G. & Scholes, K.: Exploring Corporate Strategy, Prentice Hall 1993.

Krause, F.: Knowledge integrated product modelling for design and Manufacture, The Decond Toyota Conference, Aichi Japan, 1988.

Mertz, J. & Vølund, L.: Master thesis: Product Configuration at NEG Micon A/S, Technical University of Denmark, 2000.

Mortensen, N. H., Yu, B., Skovgaard, H. and Harlou, U.: Conceptual modeling of product families in configuration projects, 2000.

Mass Customization Facing Logistics Challenges

Schenk, M. & Seelmann-Eggebert, R.

Summary: Introduced by Joseph Pine in 1993, the marketing strategy of mass customization has rapidly become indispensable for the strategic development of many enterprises. The basic idea of mass customization with all its different facets excites marketing and retail workers as well as marketing directors and CEOs. Nevertheless mass customization is still not yet a world wide standard. The reason for this can be seen in the complexity of implementing mass customization in actual existing mass or serial production. Pioneering examples often focus on newly founded enterprises or exclusively on production lines already set up. Since every product, every form of production and every logistics system have evolved individually no standard solution is or can be offered for implementation in an existing production or joint production line. Thus questions such as which product, which feature and how many features should or could be individualized remain. In addition time becomes an even more crucial factor than it has been before. Logistic systems have to be redesigned in order to face the new challenges. Long transport times have to be reduced within and between production lines. Lot size 1 in the machine does not imply lot size 1 in transport. Both information and goods have to be controlled and steered in order to be at the right place at the right time. Logistics therefore plays an essential role in mass customization.

1 Mass Customization – The Classic Approach

Classic strategies of differentiation are no longer sufficient in many industrial sectors. Globally interconnected markets offer enterprises only marginal leverage for improving their operating results. Only enterprises which manage to build individual and enduring relationships with their customers can achieve strategic advantages in the face of increasing global competition. Therefore the highest directive for modern enterprises must be to ensure consistent customer orientation, always and everywhere. Modern multimedia applications such as the Internet are spreading throughout the world and gaining popularity and can be used to enhance the power of mass customization. Individual users and business partners can satisfy their individual needs in the Internet. A mere click signs a contract. When compared to the classic approach of mass production, this new development gives rise to additional challenges: the interconnectivity between worldwide electronic data interchange and the actual production and transportation of parts and goods. Having to ship and produce individual items adds to the classic challenges, especially those concerning time prolongation and cost increments.

1.1   The Strategy

The strategy of mass customization provides a way out of the dilemma between customer demand for the highest performance on one hand and customer demand for low prices on the other hand. This strategy which was originated in and taken from marketing requires providing every customer with exactly the desired products and services. Moreover, this must be done without greatly exceeding the price of a comparable standard product. Thus the advantages of mass and serial production can be combined with those of single item production, resulting in mass or serial customization. Consumers expect outstanding products and services with all the desired features at a very reasonable price and all of this as quickly as possible. Thus in both the consumer and the capital goods sectors, the enterprise's goal must be to recognize the customer's demand faster than the customer himself does and consequently to provide the right products when the customer desires them. In this situation, mass customization offers enterprises the chance to act instead of react.

2 The Back- End's Importance for Mass Customization

Many enterprises are still little aware of the extent to which the strategy of mass customization can be applied in practice. The few examples of successful mass customizers focus on new business ideas often connected with Internet start-ups. In terms of production, the most celebrated pioneers have built up new lines where exclusively individualized products are made. Most of them made large investments and have not yet earned their return on investment (ROI) or have gone out of business. Today, however, most executive managers and decision makers recognize the necessity of being able to offer their customers first-class service. Hence, for some, mass customization appears as a welcome deliverance. This implies satisfying every customer demand and therefore frequently contributes to rapidly increasing complexity in all internal and external logistical business processes. In this context, compared with the classical meaning, the definition of logistics is enhanced. Apart from the mere transport of products and goods, this includes the steering and control of production lines, the holistic information flow within and without production and the planning of the entire supply chain. The hasty individualization of products and process structures can however lead to development past the market and thus past the customers. Before beginning far-reaching measures for restructuring and extensive investments in new machinery and IT solutions, enterprises must become aware of which form of individualization is right for them and to which extent the introduction of this concept will influence the existing process structure and logistics system.

2.1 The Demands on Service

In order to successfully introduce the concept of mass customization into a company, the entire production process needs to be reconfigured. The role of the customer acquires a new dimension and takes on greater importance than the role the customer presently holds in the classic approach to mass production. The customer now becomes the cause and consequence of the holistic process cycle. Individualization can only take place if the customer's wish has been formally expressed.

Figure 1 explains the two different ways the production process can be triggered: by transforming customer data into production data or by making inquiries of the customers. It is important to point out that the customer "pushes" the entire production process in both instances.

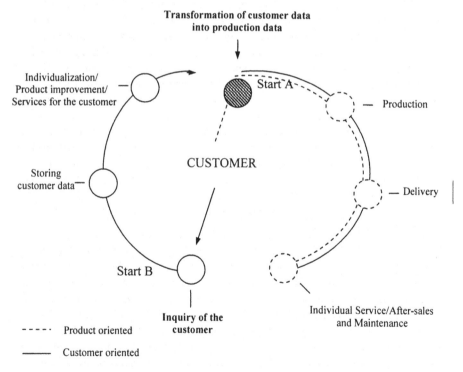

Figure 1: Product and Customer Oriented Approach to Mass Customization

2.1.1 Product Oriented Approach to Mass Customization

Start A in Figure 1 represents the beginning of a cycle where the production process is initiated by transforming customer data into production data. This course of action is easily understandable. The client specifies the product desired and the company basically adapts, configures, produces and delivers it according to the specifications provided.

Building a good relationship with the customer is vital. Assuring first-class after-sales service that suits every customer's needs is an essential approach as is offering excellent maintenance.

In this approach the information is needed merely to specify the product parameters according to the individual desires.

2.1.2 Customer Oriented Approach to Mass Customization

This procedure is different in a customer oriented approach. In Figure 1, this approach is activated in Start B. In this cycle, the customers are initially asked about their preferences. Classically, market surveys are used in several different ways. For example, mailing questionnaires to the customers is a common, though inefficient, solution, since the normal rate of returned forms is approximately 10%. Direct advertising using call centers is a more effective solution but results in higher costs and needs to be repeated frequently. Another way to ask the customer is to use direct interviews, which prove to be more efficient since they explore the customer's demands more intensively. Direct interviews require highly qualified personnel to conduct them as well as more time, which incurs higher costs when compared to the other approaches. The company can stimulate feed-back from the customer by giving the customer samples of the customized product or another kind of reward such as small presents and company souvenirs or even money. In comparison to the product oriented approach, here the company wants to learn about the customer in order to individualize service and the customer relationship. This aims at strengthening bonding with clients with the prospect of future sales opportunities.

Though other types of information about consumers' personal life and family status, for example, would be valuable, they are quite difficult to obtain, because customers are not often willing to let the company have access to their private sphere. Several global providers of individualized nutritive additives and vitamins with a business mainly based in the Internet, successfully overcame this problem by asking customers about their family and marital status. With this type of information the providers can, in a way, predict which kind of vitamins the customer's family would need, thus providing customers with customized products for the whole household.

All the information acquired concerning the customer, including orders and demands, is then stored in a database and can be used for advertising and improving individual service and can be used as an input to successfully introduce customer co-designs. This information is also used to implement Customer Relationship Management, which is essential for retaining customer loyalty but also for acquiring new customers since it increases the company's image on the market.

After the steps of Individualization, Product Improvement and Services for the Customer, the cycle depicted in Figure 1 continues to transforming the customer data into production data and afterwards to the product oriented approach just focused on.

3   Practice

The change to mass customization can often only be effected by using existing production capacities. In fact, by using highly flexible machines and tools it is easier to deal more successfully with the production time when producing lot size 1. But it requires a large investment that SMEs are often not willing to make due to the uncertainty of the outcome and to the cost efficiency. Still, it is possible to respect in most of the cases the demand of the SMEs of not increasing the current production capacities when moving into mass customization, as will be shown. Following the strategy of mass customization, different construction plans have to be generated for each order/customer and individual production information has to be distributed. The clever assortment of modern production and information technologies combined with an intelligent logistics system can thus make mass customization a strategy accessible for every company. The planning and design of the holistic value chain however most likely has to be adapted, redesigned or customized in its orientation. Employees have to be trained to understand and accept this new production method. Consistent modularization of products and manufacturing processes is strongly recommended, since this standardizes the different parts of a product. The traditional rule 'the more standardized, the lower the costs' has not ceased to exist after all. Thus classic single product or variant manufacturers in the consumer and capital goods business are offered opportunities to sustainable reduce manufacturing costs and complexity costs.

Information Complexity

Generally, two initial situations can be detected on the way to mass customization–single item production (mainly capital goods) and mass production (consumer and capital goods). In both businesses, an increased volume of data is emerging, produced by a sharply increasing frequency of recurrence of the 'New Configuration'. Classically, one order contains several products. In mass customization one order is one product for one specific customer. Thus the information's complexity is not only generated by the increase of orders controlled but also by an increase of information. This information, such as the client's name and address, needs to be handled along with each order. An even larger potential of mass customization can be utilized by introducing Customer Relationship Management (CRM). Here the client's information is stored, altered, updated and used for specific and individual client treatment and marketing as mentioned in section 2.1.2. Thus in order to cope with a large amount of data, IT systems chosen to support mass customization must be able to introduce new concepts of intelligent information logistics. Information logistics refers to the steering and control of all the information flows.

3.1 Lead Time – The Crucial Customer Variant

With mass customization, the customer expects to receive a product that exactly matches his specifications. According to market research such as by Piller et al., a customer is willing to pay 10% to 15% more than the price of a standard product in order to obtain a customized one. But a customer is definitely not willing to wait a longer time for that product. In order to meet the challenge of delivering a customized product in the same period of time of a mass produced one, the information logistics has to be coordinated quite efficiently.

In addition to the information logistics, companies need to face the classic challenge of having materials and parts in place on time. New logistical requirements arise for classic standard and mass production when mass customization is to be applied. These result from the new link between production start and receipt of the customer order. Depending of the complexity of the product there can no longer be buffer storage to keep inventory in reserve. What is more, the reduction down to production lot size 1 requires an optimization of the logistics. The impacts on the supply chain can be seen not only in the use of intelligent machines and universal control of the production flows, but also in spatial aspects. Today, consumers can buy classic mass products and take them home. For that reason, in the case of mass and serial customization, the waiting period for the end customer must be held to a minimum in order to be able to exploit the desired sales and market opportunities. Producers and suppliers are geographically 'close' to each other in an ideal way. Thus shipment processes of long duration along the supply chain and distribution times can be kept short, which decreases the lead time as well as costs.

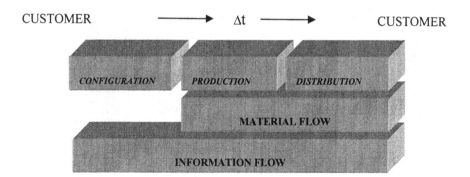

Figure 2: Material and Information Flow

Figure 2 shows the material and information flow in the several stages of the value chain throughout the production process: configuration, production and distribution. The time interval between the first phase of order placement by the customer and the last phase of customer receipt of the final customized product is represented by Δt. Reducing this time interval is crucial for the successful implementation of mass customization. This Δt should be kept as low as in a classic mass production strategy. In order to cope with such a great demand, the technological process, the logistics (process design) and the information systems must be optimized. The method for optimizing the logistics (process design) and the information systems is discussed later in this essay. Although the technology aspect is not within this essay's scope an example from kitchen manufacturers deserves mention. In some of these companies, it is possible to manufacture the furnishing for an entire kitchen in approximately two hours in accordance with the customer's specific demands. This can be done by using modern technology that automatically transforms customer data into production data.

3.2 Logistics Cycle

Before implementing mass customization, it is essential to evaluate the company's current capacity, the type of product, the flexibility in the production line and of course whether introducing such an innovating concept in the enterprise is viable. Configuration, production and logistics have to be perfectly coordinated and organized in order to achieve positive results for the company.

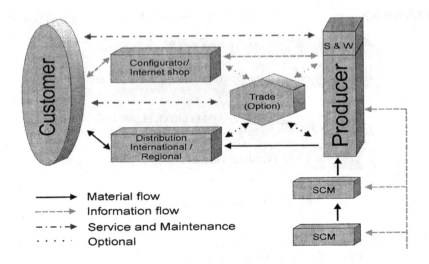

Figure 3: Model of Logistic Cycle for Mass Customization

Figure 3 shows the different links between the producer and the customer or, in other words, all the operations executed throughout the logistic cycle. It is also comprised of all service and maintenance in the after-sales period. In fact, an enterprise with a mass customization strategy needs to maintain an intensive relationship with every single customer. Not only this bonding makes it possible to retain that customer, but it can also be a useful source of information from which new ideas can be derived, leading the enterprise to more efficient service that will be rewarded by the customers.

3.2.1 Configuration

The configuration process needs to be approached differently depending on whether the particular company is an established one with a mass production strategy or a start-up,. In the former, production and distribution need to be redesigned and adapted to the new demands, while, in the latter, the configuration needs to be fully created. There is, of course, no strategy that fits every company, product or production and logistics process. Every case requires individual study and planning in order to find the most cost-effective method.

Innovative web applications in e-commerce facilitate the efficient automation of individualization in mass markets. In contrast to classic internet-shops, the user is then offered a genuine value added through individual product consulting and really appropriate products.

Nowadays, the Internet is changing the way business is done and has proven to be very valuable in implementing mass customization. This is achieved by allowing the customers to place their orders in a normal company web site. It is less time consuming, and therefore cost saving, to have the customer's specifications directly on the homepage, which are afterwards dealt with in the company itself.

The inclusion of the entire supply chain is made possible by applying internet and intranet configurations (business to business, customer to business). The end customer's data can thus be distilled and all supply chain partners receive the needed information. A frequent drawback is the difference between the data type introduced by the customer and the one existing in the company, which can also be different from the data type of all supply chain partners. Since it requires a constant conversion of data, this diversity of information causes delays in the logistics cycle as well as mistakes due to false interpretations.

Another important aspect is the amount of data the customer introduces when placing an order. This should be as extensive as possible in order to hasten the entire process. Assuming that the customer places the orders in the Internet, the Web site should include as many features as possible, but still be easily comprehended by the user.

Currently, there are several companies such as Lego and Reflect.com that have Internet based customization and that are using specific software for individual customer configuration. One of the most popular examples is the opportunity to configure some features of a car such as colors, airbags and sunroofs in the Internet. Companies such as BMW and Ford allow customers to introduce their specifications on the web, and these are then sent to a local car dealership which contacts the customer. Other specifications can be made, but not through the internet, only by contacting the actual car dealership.

When it comes to logistics, total coordination between all parts of the supply chain is crucial. This of course includes not only the information flow but also the design of how the different players are intertwined. As mentioned already, the customer is not willing to wait longer for a customized product than for a standard one. Again, the modularization of products has proven to be of great importance since it decreases the lead times for delivery.

3.2.2 Production

An approach to investigating the challenges that emerge in production when changing from mass production to mass customization is based on the model in Figure 4.

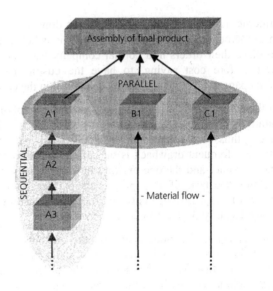

Figure 4: Sequential and Parallel Cost Centers for Mass Customization

Sequential and parallel logistical processes can be examined separately or conjointly. For the sake of simplification, the model is restricted to only three cost centers in each direction. The overall value will differ when more or fewer cost centers are investigated but the general tendencies and relations will remain the same. Exponential tendencies can therefore be detected easily. Simple mass customized structures are examined with this model. The 'production model' is used to compare the initial situation of mass or serial production with that of simple or complex mass customization.

Thus simulations are performed at that point in the same production line where an increasing number of individual parts is being produced next to the normal mass production. Overall cycle time, machine load and inventory buffer are the parameters investigated.

Simple individualizing procedures are characterized by a single individualizing step. In this case, the individualization occurs only once in the entire process. Applied to the model, this could take place in any of the cost centers marked with A1, A2, A3, B1 and C1. Locating the only individualization step just before the final assembly primarily avoids complexity costs arising from planning, labeling, individual transport, etc. In the case of an early individualization step in the process chain, the aforementioned factors might cause severe time prolongation and thus result in extreme cost increases and customers disappointed because of the possible late arrival of the product. Nevertheless, the plus in sales figures due to a 'new idea' and the exclusive market position may well equalize if not exceed the negative results. This of course depends strongly on the individual situation of the enterprise, the product and the market focus.

When multiple individualizing steps are projected, more complex structures and logistical challenges arise depending on market needs and product possibilities. The overall production time varies greatly and does not permit forecasts of delivery times for the customer. The average production time tends to grow. Some of the inconvenient conditions can be overcome using classic methods such as outsourcing or re-engineering. Individual algorithms for order management and creating transportation lots that differ from production lot size 1 can help to relieve production. In special cases, the production lot size does not necessarily have to be size 1. Some production processes allow simultaneous treatment – cleaning, bathing, painting and such - of different products. Experiments with the model show that a sequential individualization of two or more steps influences the production more than multiple parallel steps do. This influence is reflected in the increase of the process time and frequent buffers, leading production in exactly the opposite direction it was meant to head in using mass customization.

The initial model can be very easily expanded in order to reflect the situation in a supply chain. The former cost centers of the model are then seen as individual enterprises that are integrated into the holistic logistics cycle. The parameters investigated change into the material flow between enterprises, information and data exchange, the degree of cooperation between enterprises, and combined planning procedures. For individual enterprises, production cooperatives and supply chains, intelligent product planning can lead in advance to moving the point of individualization closer to the final assembly and can thus help avoid difficult and expensive logistics solutions for mass customization. Modularization of the products and processes in this context makes production in standard variants and lots possible for individual enterprises of a delivery chain. Only the product modules with individualized parameters are then affected by the increased requirements, which altogether burden the supply chain less. In the case of modular product layout, it is essential to define the interfaces of the modules exactly. Modifications, brought about by customer specifications in the module, may not violate the standards of the interfaces agreed upon.

After applying the model in Figure 4 to all possible combinations and variations, one specific trend was detected influencing the implementation of mass customization in existing production lines. Figure 5 does not scientifically represent all exact values but shows the overall increase-behavior of all dimensions. The dimensions measured [D] consisted of the overall cycle time, machine load and inventory buffer. Using structures of production lines and sequential individualizing processes, the value of the dimensions measured [D] shown in Figure 5 increased significantly when only up to 25% of all parts produced in one production line were individual.

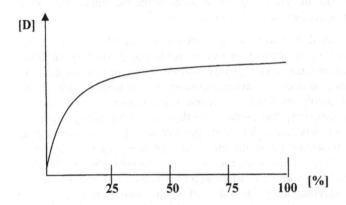

Figure 5: Increase-behavior of Product Dimension

Augmenting the number of individual products being produced in a hitherto exclusively standard production line to a net result of 50% individual products and 50% standard products results in only a slight increase in relation to the initial increase. The values measured are of course valid for individual production and the standard production conducted parallel.

That kind of behavior leads to a firm conclusion for the implementation of mass customization: running try-outs in an existing mass or serial production proves to be a complex task to accomplish. When faced by such drastic increases as in the model, an initial pilot phase seems only reasonable when a continuation of the project is seriously considered, the market possibilities have been thoroughly evaluated and the feasibility is beyond doubt.

Modularization

The standard approach to mass customization is to modularize the product. Modularization offers the advantage of creating a wide range of possible combinations and therefore results in a broad palette of choices for the customer. Depending on the product type, the selections can reach up to several million different options for the initial standardized product. Clever modularization also offers the possibility to separate standard production from customer specific production. If, for example, only one out of four modules can be altered by the customers, the remaining three can be produced using classic production methods as in Figure 6.

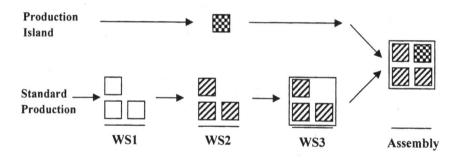

Figure 6: Modular Production

For the production of the individual module or in cases when it is not possible to implement modularization or an order cannot be fulfilled by using the standard modules, the flexibility of the production needs to be extremely high. In that case, the company must be able to quickly respond whether the request is feasible or not. Of course, declining customers' orders is not the policy of strongly competitive companies. For that reason, a different approach needs to be considered. To deal with this type of request the production line must be completely flexible in order to respond quickly and effectively. Its design should be easily changeable and the workers should have all the necessary skills to cope with the situation. Preferably, the line workers themselves should be able to make all the necessary arrangements, which would obviously reduce the lead time considerably. Individual production islands can be introduced. Parallel to standard production, they contain highly trained and flexible personnel and/or machines.

3.2.3 Distribution

Mass customization is only possible if materials flow in an expeditious manner. Synchronizing all the different parties of the supply chain is certainly quite demanding and challenging but absolutely essential if a successful mass customization strategy is to be implemented. Decisions need to be made by the minute when the objective is to shorten lead times and delivery times as much as possible.

Reconciling low levels of stock and short delivery times implies a close link to the suppliers and distributors as well as an effective flow of materials inside the company. Flexibility is again the key as is being close to the customer. Therefore outsourcing is often seen as the best choice, since fixed costs are reduced to the limit.

The need to postpone the individualization step in the mass customization strategy as much as possible implies a strong link to the outbound logistics. Third-party logistics plays a major role. Apart from coping with the aforementioned postponement and with all the distribution processes, they offer the customer services such as delivery time information and even direct marketing.

By using an outside source in logistics, Adidas, for example, managed to lower the value of its shipping costs between Asia and Europe. Costs decreased from approximately 10% to less than 3% of the final retail price of individualized sport shoes.

3.2.4   Information Systems

The synchronization of all the activities that occur between the customer's placement of an order and until receipt of the product is only possible with an efficient flow of information throughout the entire process. Personal customer data includes shipping address, payment forms and delivery specifications such as date, method (home delivery, wholesale pick-up) and individual parameters concerning a product. Upon arrival at a company, this data needs to be split in order to fulfill several tasks. Inventory stocks (raw materials) need to be checked with one's own and suppliers' capacities. When processing individual products using a parallel process structure (Figure 4), matching individual components becomes an additional challenge for production.

Radio frequency identification devices (RF-ID) can help to significantly reduce not only the amount of paper being handled but also data being processed. Transponders, as RF-ID are also called, can contain information about customized working steps and specifications together with delivery date and distribution information. They can be attached to the products or assembly carriages. A central software system no longer needs to compute or distribute information. Specific consumer data is attached to the product itself already while in production along the supply chain. The required production information can be read manually or automatically. Transponders make it possible for automated production to be steered and controlled not by a central software but by the product itself, i.e. decentralized.

When equipped with read/write abilities, transponders can furthermore serve as quality control media. Life cycle data can be stored on the RF-ID. This includes, for example, the address of the owner, maintenance dates as well as the personnel that serviced the product and service information that describes maintenance or repair steps. The amount of centralized information needed is therefore drastically reduced and an extra value is added to the product.

The information included in the product guarantees not only that it reaches the right customer but also provides data concerning all the people involved in its production process. Thus responsibility for a defect can easily be detected so that, in the event the product turns out to be defective, it is easier to find where the responsibility lies.

3.2.5 Quality Management

Although not yet focused on, the issue of quality should not be underestimated by any means or given a secondary level of importance. As discussed, the customer wants a product that fits individual specifications, wants it as quickly as possible but also expects it to be of finest quality. To deliver high quality products to the customer, the enterprise needs to follow rigorous standards in production and have high quality raw materials. Even though this does not differ much from the reality in a mass production strategy, it is much more demanding when applied to mass customization, mainly because the company cannot afford to produce defects in the latter. If a product is poorly constructed then a new one has to be produced, since each manufactured good is individual and irreplaceable. This could increase the delivery time to the customer. Therefore a zero defect policy is not only a way to reduce costs but also a necessity when it comes to mass customization.

The company may not regard quality as the final result but as a continuous process to be implemented. Following a Total Quality Management policy is quite important since it focuses on detecting potential problems before they arise. This preliminary detection avoids the need to inspect employees' work, the necessity of correcting errors and the possibility that the customer might receive a defective product. Such prevention is clearly time and cost-effective. In order to achieve good prevention of defects, the initial configuration of the production process must correspond with the quality principles the company is aiming for. Moreover the employees as well as the company's suppliers and business partners must be totally committed to achieving high standards of quality.

3.2.6 Education

Implementing a new production procedure and acquiring a different strategy in a company is a lasting and demanding process. All employees have to be engaged in achieving the defined goals. Considerable knowledge of the innovative methods is essential. For that reason education is a crucial element.

When moving from mass production to mass customization, an immediate consequence is the greater importance of the line workers' flexibility. Producing according to customer specifications implies creating unique products and consequently finding one-of-a-kind solutions. It would be quite time-consuming and expensive to thoroughly discuss which solution is the best every time. If these solutions came directly from the line worker, they could be implemented immediately.

Workers usually expect higher payment in return for acquiring other skills and for accepting different working habits. Sometimes workers do not respond in a very positive way to the new competence requirements and to the higher level of responsibility. Therefore job enlargement schemes may not be easily implemented or not even at all feasible. This initial inertia can be reduced by using interesting educational tools, for instance, virtual three-dimensional and interactive simulation programs. This kind of equipment makes it possible for workers to be in close contact with the type of environment they will work in, enabling them to participate in a wide range of hypothetical scenarios. Apart from being extremely interactive, it is also cost saving, since this kind of software is usually less expensive than real machinery. What is more, the return on investment when a machine has been used for educational purposes is lower than that when the equipment has been used in the production process itself.

The visualization techniques can also be used for research. Simulating a production line is less time consuming and more cost effective than building one. Features can be easily changed and/or added without altering the current production line, thus resulting in not having to stop or delay the production process.

Not only do the line workers have to be fully aware of the new strategy but every one inside the company has to be fully committed in order to implement such an innovative technique. Special business games can help to point out the demands of a mass customization strategy to the entire staff of a company or even a supply chain. These educational tools are employed to create a general awareness of the upcoming challenges. By solving cases the staff involved understands and better supports future changes that might occur in their own process structure.

The customer no longer plays a passive role in the production process. The customer's participation is absolutely required and most welcome. In order to make individual specifications, the customer has to be aware of all the possible choices available. Therefore great emphasis has to be placed on the information provided to the customers.

Sometimes some special specifications can be impossible to fulfill, this means that the customer must have the information about what not only can be but also what cannot be chosen. Therefore, education of sales personnel has to be emphasized, since they must be able to provide the client with all the information necessary, not only about the product itself but also about the features of the production process. By knowing how this process works, the sales personnel can immediately verify if a certain request is viable or not.

Conclusion

In 1759 Samuel Johnson remarked, "The trade of advertising is now so near to perfection that it is not easy to propose any improvement".

It is still not easy to introduce improvements but one can hardly say that perfection has been reached. Several developments must be made in order to reach every specific individual rather than the population at large.

On account of diverse reservations and a lack of local examples, mass customization is still not very widely spread. Nevertheless the discussion about this subject is strongly increasing. This is happening against the background that mass customization is regarded globally as one of the best strategies to maintain and consolidate enterprises market positions in the future.

Mass customization is definitely the right approach to reach the much craved for market share. Just as they always did, customers know what they want but now they are not willing to settle for the second best choice. The companies that provide them with the particular product or service sought will succeed and mass customization is the only way to achieve this in a cost effective way.

This new approach will change not only the production process, but the entire way the supply chain is organized and coordinated. Correctly applied, logistics becomes the key enabler for mass customization in an enterprise and thus the factor for sustainable market growth.

References

Duray, Rebecca, Peter T. Ward, Glenn W. Milligan, William L. Berry:
Approaches to mass customization: configurations and empirical validation.
Journal of Operations Management, 2000.

Gooley, Toby B.: Mass customization: How logistics makes it happen. Logistics
Management and Distribution Report, April 1998.

Hibbard, Justin: Assembly Online. Information Week, April 1999.

van Hoek, Remko I.: The Role of Third-Party Logistics Providers in Mass Customization.
The International Journal of Logistics Management, Vol. 11, No. 1, 2000

Lavidge, Robert: Mass customization Is Not an Oxy-Moron. Journal of Advertising
Research, July-August, 1999.

Pine II, Joseph B. and James H. Gilmore: The four faces of mass customization.
Harvard Business Review, 1997, 91-101.

Piller, Frank T.: Kundenindividuelle Massenproduktion. Hanser Verlag, 1998.

Seelmann-Eggebert, Ralph: Dienstleistungsaspekte der Mass Customization. Industrie
Management, April 2001, 69-72.

Seelmann-Eggebert, Ralph: Mass Customization facing logistics challenges. International
ICSC Congress - Australia, Vol. 1, Dec. 2000, 174-177.

From Mass Production to Mass Customization: Impact on Integrated Supply Chains

Caddy, I., Helou, M. & Callan, J.

Summary: 'Mass Customization' can be considered the 'holy grail' of manufacturing in which products made are perfectly attuned to customers' wants and needs. This paper analyses what should be seen as mass customization by developing the mass production/mass customization continuum. In addition, whether all products can be easily mass customized is debatable; where products consist of combinations of goods and services, organizations may find that one component or dimension of the product is more easily customized. The main issue discussed is supply chain change due to mass customization of products. Given that mass customization is significantly different to the mass production paradigm, there should be changes in either the way the supply chain is configured or in the way the supply chain needs to be managed. This paper will consider what changes may impact on a supply chain for mass customized products using the generic supply chain framework. The analysis concludes that both supply chains and their management will be different to supply chains aligned to more traditional manufacturing processes.

Keywords: Mass customization, Supply chains, Supply chain management, Generic supply chain framework

1 Introduction

Supply chain management has focussed predominantly on traditional business operations and processes. For example, Das and Narasimhan (2000) discuss manufacturing purchasing competence, but this research makes an assumption this paper will try to avoid. Manufacturing process structure does not change, but merely process location (in an organization sense). A lot of similar research has been conducted presuming that manufacturing methods do not change. There is only a need to reassess who can and who will perform the process. Indeed little has yet to be offered concerning supply chains in either the services sector (Beldona, 1999) or the non-profit sector (Helou and Caddy B forthcoming paper). In most cases, traditional mass production manufacturing processes are taken to be the norm, while other forms of manufacture the exception. This paper rejects this presumption by examining mass customization. It will consider whether changes to manufacturing process structure impart concomitant changes to the supply chain, or supply chain management. In moving from mass production to mass customization, how radical could these changes be? Considering the motor vehicle industry, on which too much research activity is focused, there are two radically different scenarios that could emerge due to adoption of mass customization technology.

First, further disintermediation on the supply chain's customer side could occur. Motor vehicle manufacturers who perceive their core competencies as firmly embedded in the manufacturing process, coupled with a strategic need to dominate their markets, will invest in mass customization technology and adapt the way motor vehicles are made. However, mass customization needs complete and accurate customer information, which is easier to obtain directly from customers than through an intermediary. In fact customers may never enter a car yard again, but conduct search activities, go for virtual test drives, and make final purchase decisions through an Internet portal. General Motors (gmbuypower.com), and BMW (bmw.com/bmwe/products/individual) provide a glimpse of what these portals will look like. Although these sites do not provide features such as virtual test-drives this limitation would be removed with implementation of projects such as Internet2, Abilene, or Dante (Donegan, 2000; Wright, 2000; Berendt, 1999; Lange, 1999). Croom (2000), in discussing the use of such e-commerce facilities, concludes that not only is completeness and accuracy of information transfer improved, but there are significant cost reductions in obtaining and then using the information, giving incentives for motor vehicle manufacturers to switch to mass customization. The other scenario involves reintermediation on the supply chain's supplier side. Motor vehicle manufacturers whose core competence is car design and servicing of niche markets could transform themselves into "car design houses". Car designing and creation of customization manufacturing software become their new core competencies. Motor vehicle manufacture is performed by strategic suppliers located in appropriate geographic regions close to customer concentrations (themselves are strategically related to the car designers and not the car manufacturers). The supply chain now incorporates a strategically important virtual supply pipe.

Accordingly there is support that mass customization can alter supply chains. This research will analyze the types of changes using the generic supply chain framework (Caddy and Helou, 1999) as the research instrument. A by-product of this analysis is confirmation of the framework's robustness given its application to non-traditional manufacturing. Further research arises through the implications of the analysis presented. Applied research will use the generic supply chain framework within non-manufacturing industries as well as to other manufacturing technologies, such as mass customization.

2 Supply Chain Management Issues

Given a move to mass customization, the organization would then need to make another, and dependent, strategic decision. Should the organization acquire the necessary mass customization technology itself, or should it outsource product manufacture to others? As Collins and Bechler (1999) have stated (along with many others), outsourcing has been seen as a viable alternative to the organization performing a business process itself. A recent survey reported on by Trent and Monczka (1998) indicated that not only would organizations rely more on third parties to produce product, but that organizations would also rely on third parties in the use of process technology. Although not discussed by Trent and Monczka (1998), one form of process technology is mass customization. Many organizations have outsourced their IT function, believing that IT service delivery is not a core business function (Applegate, et al., 1996; Behara, et al., 1995; Rao, 1995; Loh and Venkatraman, 1992). However, the viability of IT outsourcing, has also been questioned (Langford, 1997; Currie, 1996; Lacity, et al., 1996; Lacity and Hirschheim, 1995). Issues for organizations to think about carefully when looking at outsourcing are long-term overall costs and organization capture. Is there a cost saving over the life of the current technology? Is the organization transferring expertise and knowledge to the outsourcer that it will find difficult or almost impossible to regain at a later stage? Not all organizations find themselves in the same position and so some may decide to outsource while others do not: hopefully both do so for good, sound strategic reasons. Two additional factors critical to successful outsourcing should also be considered.

First, there must be an ability to identify appropriate suppliers and then build effective relationships with these suppliers. How good is the organization at doing this? Given significant levels of investment in mass customization technology, the participants will be looking for long-term agreements rather than short-term. In these circumstances, as Collins and Bechler (1999) state, suppliers are more partners, and trust is a vital factor in developing successful long-term relationships. Collins, et al. (1997) developed two models to assist organizations considering whether to outsource or not, viz. the Supplier Profile Matrix and the Stairs of Transformation models. These models help managers understand the difficulties and complexities associated with partner selection, and then building and nurturing the relationship once the outsourcing decision is made. Second, outsourcing arrangements are never static and so need to be managed accordingly. Despite good management, some outsourcing arrangements will sour, in which the partners find increasing levels of disagreement and dysfunction between themselves. Where this occurs, a critical issue becomes the ability to extract from, or substantially revise and rejuvenate, the arrangement. Collins and Bechler (1999), in discussing Skoda's experience make a very relevant point: the greater the level of commitment by the participants then the more difficult it will be for those participants to terminate their outsourcing arrangements. Ease of separation also depends on what happens at the beginning of the arrangement. However, once joint investments are made and there is development of operational inter-dependence, then termination becomes far more difficult.

Outsourcing mass customization would be different to that discussed above. First, until mass customization technology becomes the rule rather than the exception, organizations will have little choice of outsourcing partners. Second, a greater level of knowledge is required to use mass customization technology effectively. Accordingly, there will need to be greater care and diligence exerted by organizations in identifying potential partners and greater effort required in managing the relationship downstream. Given these factors the difficulty in terminating these sorts of agreements will be high from commencement rather than with maturity as compared to traditional manufacturing outsourcing. Another factor to assess when deciding on outsourcing or not is the ease of mass customizing the organization's products.

3   Mass Customization Issues

3.1   'Ease' of Mass Customization

In applying mass customization to producing goods, providing services, or supplying products (goods combined with services or vice versa), there will be different levels of difficulty across different industries. For example, will true mass customization occur with respect to motor vehicles? That is, the customer can choose colour, engine size, number of cylinders, steering wheel shape, car seat sizes and shapes (all of which could be different) and so on? Alternatively, is this level of mass customization infeasible? That is, the capital investment required would push the final product's price well out of most customers' reach. If so, then motor vehicle manufacturers may decide on a method of product delivery termed 'mass optioning', i.e. a large and quite diverse range of standard options are on offer to the customer. Indeed 'mass optioning' can give the appearance of mass customization as the number of permutations in option selection will often be greater than the size of the potential customer population. As indicated in Figure 1 below, a 'mass optioning' strategy is a long way removed from the mass production paradigm that has been dominant for most of the car manufacturing history. Furthermore, an enhanced level of mass optioning for motor vehicles, extending the current range of motor vehicle components that can be 'optioned', is not beyond the realms of the possible in the future. On the other hand with respect to glass manufacture, the absence of multiple inter-linked components that need to be assembled should make mass customization an easier proposition. This would mean that manufacture would go beyond mass optioning, such as embossing a monogram or some other family emblem on the face of each glass. It would not take a great leap forward in technology for customers to be able to provide designs of glassware themselves, select and modify template designs, or merge different templates to produce a unique product designed to satisfy their individual taste.

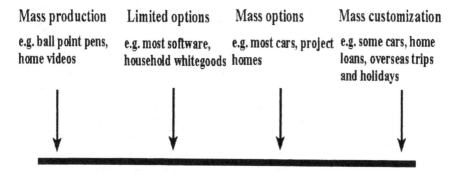

Figure 1: The Mass Production/Mass Customization Continuum

3.2   Mass Customization Versus 'Mass Optioning'

As indicated by Figure 1 above, the choice between mass production and mass customization is not an either or decision. Rather products fall along a continuum in which mass production and mass customization represent the two extremes. The choice of how much or how little customization of a product should occur will ultimately depend on customer preference for a standard product compared to something tailored to their unique wants and needs. In the latter case, the question becomes whether customers can communicate their wants and needs. For example, would customers have the engineering expertise to custom design a car from the ground up, or would they be more comfortable in selecting a standard model of car and then select from a set (which can be a large set) of predetermined options? Another question to ask: do customers want mass customization? There are indications of such a trend. Take for example the humble toothbrush, an exemplar in earlier years of mass production. In recent times toothbrushes now come in different shapes, the bristles of the brush have many different configurations, some toothbrushes indicate the amount of life they have left, and others come with removable heads than can be replaced with new ones. On the continuum shown above the toothbrush has moved from mass production to somewhere between production with limited options and production with mass options. Finally, as indicated by the mass customization examples, organizations may find it easier to customize services rather than goods. Where products combine goods and services then the organization may find it easier to customize the service rather than the good component.

4 The Generic Supply Chain Framework

This framework describes the factors that impact on an organization's supply chain and the interactions between these factors (Caddy and Helou, 1999). The "Organization Strategy/ Structure" factor is looks at issues such as number of supply chain links, with either immediate (or first tier) suppliers or links between a first tier supplier and its suppliers. Links can be simple dyads, or more complicated networks. Supply chain analysis from this view is concerned with the contingencies that affect its topology. For example, Hall (1999) states there are differences between 'first tier' suppliers and 'lower tier' suppliers. Concepts such as 'lean' organizations and 'agile' supply chains fall under this factor (Barlas, *et al.*, 1999; Naylor, *et al.*, 1999). Handfield and Nichols (1999) claim supply chains include internal functions, upstream suppliers and downstream suppliers. Internal functions transform inputs from suppliers into final products. Upstream supply chain management ensures a smooth flow of materials. Downstream supply chain management ensures completion of all processes a final product passes through on its way to the customer. Although supply chain topology is important, it is not the only consideration. Mariotti (1999) claimed "the weakest link in the most advanced supply chains is not technology, not software or hardware, but people - or rather the level of trust between people who must cooperate and collaborate to get results". Burnell (1999) stated senior management commitment was critical in effective supply chain development and management. Handfield and Nichols (1999) state that poor human relationships in any supply chain link may have disastrous consequences for the supply chain, irrespective of supply chain topology or the strategic intent of organizations involved. The information technology (IT) factor considers how advances such as e-commerce (Harrington, 1999; Hickey, 1999; Hickins, 1999, Thomas, 1999) will affect supply chain operation and management. One key issue is integrating traditional information systems to the supply chain (Michel, 1998; Jenson and Johnson, 1999). Another issue concerns the information and the IT infrastructure capability to allow information sharing. While in general greater information sharing is seen as beneficial (Kiely, 1998; Gavirneni, *et al.*, 1999) not all researchers agree (Singer, 1999; Trunich, 1999). In terms of interactions, greater information sharing (IT view) can only occur where higher levels of trust exist between the supply chain participants (human factors view). While IT is an important component it would be wrong to regard the supply chain as just another information system.

This is one of the major contributions of the generic supply chain framework. While each of these factors is relevant to understanding, analyzing and managing supply chains, none could be said to be all-encompassing, and none could be said to be superior to the others in trying to understand supply chains. Being an amalgam of these three views, the model incorporates the benefits possessed by previous and more specific models. Moreover, and most importantly, the generic supply chain framework allows a greater insight into the development, implementation and management of supply chains by considering the effect of the interactions between these three views of the supply chain. In applying this framework to considering the impact of mass customization (or in its lesser form - mass optioning) on the supply chain, most attention will be placed on the changes likely in the interactions of these three views.

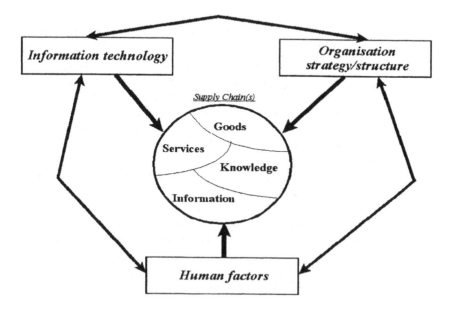

Figure 2: The Generic Supply Chain Framework

5 Applying Generic Framework to Mass Customization

5.1 Changes in the Supply Chain 'Pipe'

Mass customization technology requires more information concerning customers' preferences on characteristics of the goods, services, or products provided by the organization. That is, the impact of mass customization on the 'supply pipe' will be a greater emphasis on information as well as knowledge. On the one hand organizations will need greater information regarding customer preferences concerning the alterations either to a standard product or to product creation from scratch. On the other hand customers will require greater knowledge about the product, such as how different options interact in terms of a particular custom design, and other issues such as knowledge of the product's design limitations.

5.2 Changes in Organization Strategy/ Structure-Driven Factors

Mass customization will require organizations to change supply chain structures that allow greater knowledge transfer by the organization to their suppliers. This knowledge transfer includes knowledge about product manufacture processes, more access to proprietary customer information, both overall and in particular, such as customer preferences for product characteristics. This will be the case irrespective of whether the organization decides to outsource manufacture of mass customized products or not. In the case where the organization invests in mass customization technology, suppliers will still need greater knowledge about the components they supply and how they integrate into the final product. Greater information about end customers' preferences will be needed by the organization's suppliers, so they in turn can manage their own supply chains. Whereas in traditional mass manufacture supply chains the customer was 'hidden' from the organization's suppliers this will no longer be a viable strategy. For these reasons one major change in the supply chain structure will be a need to place a higher level of trust in the partnerships required to produce mass customized products. Given this higher level of trust, supply chain structure may have more focus on first tier suppliers to provide the organization's inputs rather than secondary suppliers.

5.3 Changes in Human Factors

For the human factor view there should be higher trust levels and more focus on managing relationships. The supply chain will require people who are better communicators and who display greater adaptability to changing circumstances, such as shifts in customer tastes. Thus organization strategy and structure interacts with human factors as indicated in the generic framework. There will also need to be greater senior management commitment to the new technology as suppliers and customers adapt to, and become comfortable with, their new selling and purchasing environments. A mass customization supply chain will be driven to a far greater degree by the customer than would be the case with other supply chains: the adage "the customer is always right" will never ring more true. Marketing strategies will need to be more in tune with customers' preferences and with recognizing changes to customers' preferences.

5.4 Changes in IT-Driven Factors

This factor is considered the one that will change the most. Mass customization technology is dependent upon IT. IT is also a necessary component for the successful development of either mass optioned or mass customized products. For example given that service-based products are predominantly developed through, and delivered by, IT technology, the mass customization of these types of products will be dependent upon the successful implementation and operation of IT. This will also be true where products are primarily goods, such as motor vehicles. One difference between goods-based products and service-based products is that it will be easier to mass customize service-based products. Indeed for goods-based products it may be the service component that is mass customized rather the goods component. For example, a motor vehicle has both a goods component (the car itself) and a service component (after sales servicing and warranty plans). Mass customization may be easier to achieve in after sales service strategies or through different warranty plan designs than it is with the motor vehicle. Therefore a viable strategy that motor vehicle manufacturers could pursue is to mass option the product's good coupled with the mass customization of the product's services.

Another issue concerns knowledge and information transfer between customer and organization (or the organization's supplier). What sort of interface should be provided to enhance the customer's capacity to customize the product? This interface will need to do two things: possess knowledge so that it can effectively interact with a customer who has less than the requisite knowledge level needed to custom design a product. In a mass production context the presumption of an equal level of customer knowledge was not a requirement to ensure a sale. However, in custom designing (or mass optioning) a product such as presumption is untenable. For example, in custom designing (or mass optioning) a motor vehicle there would need to be an accurate assessment of what computer-based facilities would need to be provided to customers so they can make intelligent and safe engineering decisions at every step in the design process. Customers should be warned interactively where options selected are incompatible or the combination is not optimal in terms of the product's safety. Mass customization (or mass optioning) cannot be a more difficult process for the customer than it is today in terms of buying mass produced products or mass-produced products that have limited options. If the customer has to obtain university level engineering knowledge before they can customize their motor vehicle then the only customers that organization will have will be engineers.

The customer interface would most likely be a Web-based interface but other interfaces could be developed. For example, the organization could build self-learning intelligent agents (Cheong, 1996). These agents would be provided to customers free of charge (similar to Internet browsers) so that the customer does not need to log on to a particular Web page. The agent would possess intelligence to ensure that customers can make appropriate design decisions. The agent should be self-learning in terms of customer preferences, and so itself is a mass customized product. This knowledge could also assist organizations to plan new products or marketing strategies to win new customers.

6   Conclusion and Future Research

This paper demonstrates that the generic supply chain framework can be applied to analyzing supply chains and their management within non-traditional methods of manufacture. The robustness of the generic supply chain framework is also proved indicating it as the instrument of choice when analyzing supply chains. As alluded to earlier, the paper also represents a beginning rather than an end, and has possibly posed more questions than it did in finding answers. Future research will attempt to find answers to the following questions:

a.   To what extent do customers use mass optioning when purchasing either a motor vehicle or a project home?

b.   To what extent does each view within the generic supply chain framework participate across different industries and between different manufacturing technologies?

c.   Do the interactions between organization strategy/structure, human factors, and information technology differ across different industries or between different manufacturing technologies, or are the interactions similar?

References

Applegate, L., McFarlan, F.W. and McKenney, J.L. (1996): Corporate Information Systems: The Issues Facing Senior Executives, Richard D. Irwin Inc., London.

Barlas, S., Curatola, A., Randall, R. and Williams, K. (1999): Cost management of supply chains, Management Accounting, 80(12), p. 23.

Behara, R.S., Gundersen, D.E. and Capozzoli, E.A. (1995): Trends in information systems outsourcing, International Journal of Purchasing and Materials Management, 31(2), Spring, pp.46-51.

Beldona, S. (1999); Supply Chains B A Perspective on Information Sharing in the Hospitality Industry, Conference Proceedings, Second International Conference on Managing Enterprises, Newcastle, 17-20 November, pp. 188-195.

Berendt, A. (1999). Internet2: Tapping the academic test bed, Telecommunications (International Edition), 33(7), July, p. 76.

Burnell, J. (1999): Change management the key to supply chain management success, Automatic I.D. News, 15(4), pp. 40-41.

Caddy, I. and Helou, M. (1999): Supply Chain and Supply Chain Management: Towards a Theoretical Foundation, Conference Proceedings, Second International Conference on Managing Enterprises, Newcastle, 17-20 November, pp. 112-123.

Cheong, F-C. (1996): Internet Agents: Spiders, Wanderers, Brokers and Bots, New Riders Publishing, Indianapolis, Indiana.

Collins, R. and Bechler, K. (1999): Outsourcing in the chemical and automotive industries: Choice or competitive imperative?, Journal of Supply Chain Management, Fall, 35(4), pp. 4-11.

Collins, R., K. Bechler and S. Pires. (1997): Outsourcing in the Automotive Industry: From JIT to Modular Consortia, European Management Journal, 15(5), October, pp. 498-508.

Croom, S.R. (2000): The impact of Web-based procurement on the management of operating resources supply, Journal of Supply Chain Management, 36(1), Winter, pp. 4-13.

Currie, W.L. (1996): Outsourcing in the private and public sectors: an unpredictable IT Strategy, European Journal of Information Systems, 4, pp. 226-236.

Das, A. and Narasimhan, R. (2000): Purchasing competence and its relationship with manufacturing performance, Journal of Supply Chain Management, 36(2), Spring 2000, pp. 17-28.

Donegan, M.P. (2000): Back in the lab, Telecommunications, 34(5), May, pp. 46-50.

Dyer, J.H., Cho, D.S. and Chu, W. (1998): Strategic Supplier Segmentation, California Management Review, 40(1), Winter, pp. 57-77.

Gadde, L.E. and H. Hakansson. (1994): The Changing Role of Purchasing: Reconsidering Three Strategic Issues, European Journal of Purchasing and Supply Management, 1(1), pp. 27-35.

Gavirneni, S., Kapuscinski, R. and Tayur, S. (1999): Value of information of capacitated supply chains, Management Science, 45(1), pp. 16-24.

Germain, R. and C. Droge. (1998): The Context, Organizational Design, and Performance of JIT Buying vs. Non-JIT Buying Firms, International Journal of Purchasing and Materials Management, 34(2), Spring, pp. 12-18.

Hall, R. (1999): Rearranging risks and rewards in a supply chain, Journal of General Management, 24(3), pp. 22-32.

Handfield, R.B. and Nichols, E.L. (1999): Introduction To Supply Chain Management, New Jersey, Prentice Hall, Inc.

Harrington, L. H. (1999): Using the Net to stay competitive, Transportation & Distribution, 40(5), pp.23-26.

Hickey, K. (1999): Java chain, Traffic World, 258(9), p. 46.

Hickins, M. (1999): It's an e-buyer's market, Management Review, 88(6); p. 6.

Jenson, R.L. and Johnson, I.R. (1999): The enterprise resource planning system as a strategic solution, Information Strategy: the Executive's Journal, 15(4), Summer, pp. 28-33.

Kiely, D.A. (1998): Synchronizing supply chain operations with consumer demand using customer data, Journal of Business Forecasting, 17(4), Winter, pp. 3-9.

Lacity, M.C. and Hirschheim, R. (1995): Beyond the Information Systems Outsourcing Bandwagon, Wiley, Chichester, England.

Lacity, M.C., Willcocks, L.P. and Feeny, D.F. (1996): The Value of Selective IT Sourcing, Sloan Management Review, Spring, pp.13-25.

Landeros, R. and R.M. Monczka. (1989): Cooperative Buyer-Seller Relationships and a Firm's Competitive Posture, Journal of Purchasing and Materials Management, 25(3), Fall, pp. 9-18.

Lange, L. (1999): Technology 1999 analysis & forecast: The Internet, IEEE Spectrum, 36(1), January, pp. 35-40.

Langford, R. (1997): Outsourcing: Can a business core be outside?, Business Review Weekly, September, pp. 92-93.

Lascelles, D.M. and B.G. Dale. (1990): Examining the Barriers to Supplier Development, International Journal of Quality & Reliability Management, 7(2), pp. 46-56.

Loh, L. and Venkatraman, N. (1992): Determinants of information technology outsourcing, Journal of Management Information Systems, 9(1), Summer, pp. 7-24.

Mariotti, J. (1999): Plenty of technology, but a shortage of trust, Industry Week, 248(11), p.128.

Michel, R. (1998): SAP's new view, Manufacturing Systems, 16(11), pp. 12-13.

Miller, J.G. and A.V. Roth. (1994): A Taxonomy of Manufacturing Strategies, Management Science, 40(3), pp.285-304.

Naylor, J.B., Naim, M.M. and Berry, D. (1999): Leagility: Integrating the lean and agile manufacturing paradigms in the total supply chain. International Journal of Production Economics, 62(1,2), pp.107-118.

Rao, S.S. (1995): When in doubt, outsource, Financial World, 164(25), pp. 77-78.

Singer, T. (1999): Sharer beware, Inc., 21(4 (Inc. Technology Supplement)), pp. 38-48.

Thomas, J. (1999): Thoughts on falling behind, Logistics Management & Distribution Report, 38(6), p.96.

Trunick, P.A. (1999): Loose lips sink shipments, Transportation & Distribution, 40(3), pp.65-68.

Wright, R. (2000): Will we ever log off?, Time, 155(7), Feb 21, pp. 56-58.

Key Value Attributes in Mass Customization

MacCarthy, B. L., Brabazon, P. G. & Bramham, J.

Summary: The chapter introduces the concept of product key value attributes (KVA) for understanding Mass Customization (MC). The customization potential and desirability of product attributes may be explained in this light. The concept provides a basis to understand the spectrum of MC strategies and the challenges that MC poses operationally. A number of conceptual models are introduced to explore both customization potential and desirability with respect to the customer and the producer. The challenges that MC poses for the design of effective operational systems and for strategy formulation are identified in the context of KVAs. A classification of different types of key value attributes is described and their operational implications discussed. The ideas are examined in the context of the market environment in which MC takes place. The impact of proliferation of product variety is highlighted. The nature of the customer is considered and the need to include business customers demanding product differentiation is noted.

Keywords: Key Value Attributes, Mass Customization, Conceptual Models, Operational Systems.

1 Introduction

Mass Customization (MC) attempts to provide *customized* products for individual customers without losing the many benefits of *mass* production - high productivity, low costs, consistent quality and fast response. However the definition and true meaning of MC continues to be discussed (Pilkington & Chong 2001) and its feasibility remains in question, as pointed out by a number of prominent critics (e.g. Zipkin 2001, Agrawal et al. 2001, Womack 1993). A fresh view of MC, as set out here, can help move forward the debate and spur the development of operations principles and approaches.

In this chapter we advocate a view of MC based on product key value attributes in order to understand the MC spectrum and the challenges it poses operationally. First we examine the context of MC in the market environment with respect to two elements – proliferation of product variety and the nature of the customer. We then examine the nature of customization in relation to MC. A value chain perspective is introduced for assessing value to the customer. The wider issues of value are then explored with the aid of conceptual models. These models are reviewed in the light of current trends and commercial reality. The challenges that mass customization poses for the design of effective operational systems and formulation of strategy are discussed.

2 The Market Environment

2.1 Product variety

Although hard facts and figures are difficult to divine, it is clear that many companies have to deliver more variety to compete effectively in their market places. There has been a significant increase in product variety in most product categories over the last quarter century (Cox and Alm 1998, Funke 2000). Trends in the automotive sector are not as clear-cut. Importantly, it depends on how variety is measured. Holweg and Greenwood (2000) argue that there has not been a general trend towards increasing or decreasing variety in the automotive sector when viewed in terms of available options. However, the trend of increased levels of new product introduction continues apace (The Economist 2001) In the SUV (sport-utility vehicles) category alone there are now over 70 models on sale in the US market. Not surprisingly market potential and lifecycle of new automotive models are reducing. The Economist (2001) reports that the best selling model for GM in the 1960's could achieve annual sales of a million or more units whereas today its top selling passenger car can manage just 300k units. In many vehicle categories the market is being 'sliced and diced' into ever smaller segments. Thus, the current operations strategy of many automotive manufacturers for more flexible plants is not surprising, if somewhat belated.

These factors are not just limited to consumer, grocery or commodity products. In the steel sector for instance, Coates (1998) notes the pressures in contemporary steel production with the number of grades and varieties increasing significantly whilst batch sizes and lead times have decreased in what is now a highly competitive global market.

Why is variety important in the MC context? Adoption of an MC strategy implies large increases in both simultaneous and sequential variety. The effects are felt not just within core operations but across the extended enterprise. Responsiveness, quality and cost containment are key features of MC systems that must not suffer with step changes in variety. The cost benefits of high volume, low variety manufacturing are well known and underlie traditional manufacturing strategies (Hayes and Wheelright 1984, Stalk 1988, Kekre and Srinivasan 1990). Increasing product variety potentially impacts many aspects of the operations of a manufacturing enterprise. Evidence varies on the precise influence. A survey conducted by Yeh and Chu (1991) across flow shops, continuous production and project manufacturing sought to rate the direct and indirect impacts of variety. The highest impacts were found to be on product flexibility, raw materials and changeover. Whilst Stalk (1988) believes that halving variety can increase productivity by 30%, an empirical study of the automotive sector found that only parts complexity - variety partially driven by consumer choice - had a significant impact on operations (MacDuffie et al. 1996). Other measures of variety including model mix complexity, option content and option variability had no significant effects when working within the designed envelope of variety. Fisher and Ittner (1999) studied the impact of variety in automotive assembly using empirical data and simulation and concluded that increasing the day to day variety spread had negative impacts on almost all production performance measures. Random variation was identified as being more pernicious than high option content levels.

What is clear is that variety within a manufacturing plant has undesirable effects potentially on costs, quality and responsiveness. Just as importantly, when the impact of variety is considered in the context of the extended enterprise – supplier networks, production networks, logistics and distribution systems - the potential ramifications are of even greater significance. Whether customers and consumers are demanding this variety or whether it is producer led is, in some respects, irrelevant. It is clear that customers and consumers expect variety and that producers must have effective customer-focused variety strategies to survive. However and wherever variety explosion occurs, it is necessary to understand its impact, how to live with it and possibly how to exploit it. This issue poses specific challenges for MC strategies. Of key importance is the need to focus on relevant variety.

2.2   The nature of the customer

Currently a lot of the variety experienced by producers in consumer products such as bicycles or mobile phones, is not driven directly by the individual consumer but by business customers such as retailers, stockists, dealers, distributors, catalogue merchants etc. Such customers demand fast response product differentiation that is customer-focused. Mass merchants (McDermott and O'Connor 1995) have significant power in marketplaces and may not be satisfied with selections from variants. They demand product differentiation that is distinctive but which may often be difficult to achieve within current capabilities. In a number of sectors the need for product differentiation has strong functionality or physical constraints e.g. specialised vehicles such as ambulances. Some mass producers of commodity products face such issues increasingly when responding to the need for product promotions. These may be driven by the large retailers and the mass merchants or by marketing departments within their own organisations. Manufacturing enterprises and their suppliers need to be able to deal effectively and efficiently with such a drive for 'product differentiation'.

How does product differentiation differ from a narrow interpretation of MC driven by the desires of an individual consumer? Differentiation can be viewed as a form of customization. Quality and cost imperatives of MC apply equally within this B2B product differentiation environment. Rapid product visualisation and prototyping, as well as rapid engineering and costing are also features of this environment. Undoubtedly batch sizes of one do not prevail but, as noted previously the trend for batch size reduction is being experienced across many sectors with the implication of much less predictability in demand for specific variants. The capabilities necessary to differentiate products rapidly enables companies to respond to a wider range of business customers. It can be argued therefore that rapid response product differentiation for this type of market is a limited form of MC and one that may enable the competencies to be developed for true niche production and ultimately for batches of one.

2.3 Customization in the context of MC

Product customization appears at first glance to be a relatively simple concept – the customer specifies exactly what they want and the manufacturer produces it. Simplistic views of MC see product customization essentially in this light, implying a digitisable, made to measure product that is customized on demand. However a little more consideration shows that such a 'bells and whistles at the customer's whim' view of customization is limited in the context of MC and does not mirror reality or trends in many sectors.

Deeper consideration is needed to define and interpret customization and some of its apparent contradictions. How do we distinguish between a simple colour choice for a product such as a vehicle, a fully specified engineered product such as a paper making machine, a customer-configured product such as a PC, or a product differentiated by different levels of service warranty such as a photocopier? What about a customer who wants simplicity or affordability? Removing features and functionality may be very valuable to specific customers. All of these must be seen as legitimate forms of customization.

2.4 Continually evolving customization potential

Both customization potential and customization desirability are affected by the product lifecycle. In fact what may be customizable in a product and when it is desirable to offer customization is affected by many inter-related product lifecycle factors:

- *Tastes and fashion*: consumer desires and tastes change with time. Product attributes that were subject to limited variety may require more consumer focus with the provision of greater variety. As well as personal products many consumer durables are subject to such influences.

- *Different markets*: a desirable customizable attribute in one market may be less important in another. Products are often introduced to different markets at different times and markets may follow one another. Different or changing demographic patterns may impact customizable attribute desirability.

– *The competitive environment*: what competitors are offering in terms of customization will be an important driver in staying competitive and in developing competitive advantage.

– *Production technological capability and product innovation:* Technical or operational difficulties resulting from constraining technology often limit customization potential. The Benneton example is a classic case where re-engineering of the manufacturing and colouring process enabled a greater level of customer focused colour variety to be delivered quickly but also limited the garments to a single shade. New technologies, for instance colour application technologies, can radically alter the customization potential and desirability in a product.

– *Product maturity*: what is customizable early in a product lifecycle may become standard later as costs go down and new entrants make the market more competitive. Customer expectation is also an issue. As a product type becomes mature then certain features are expected. The producer may want to hold on to a customizable feature for as long as possible, particularly if it is a distinctive capability, but changes in the price value relationship may make customization on that feature superfluous.

The UK car market provides an interesting example of evolving customization potential. In the early sixties a heater was an option. In the early seventies a radio was option. Up until relatively recently air conditioning was an option available only on some premium models but one that is increasingly being fitted as standard to family saloon and executive cars. The personal computer is perhaps the best example of evolving customer variety. The PC is in many respects a standard product. However a very large number of configurations are sold by most suppliers or manufacturers. Some such as Dell allow almost infinite variety to be specified by the customer. However the dimensions along which customization is available are continually evolving. Thus a standard PC modem is no longer offered in 14/32 and 56 Kbits/sec – almost all are sold at the highest specification. However, PC screens are still customizable by size and technology to suit the purchaser's application and budget.

Significant but continually evolving customizable variety around a common platform is a feature of many sectors, particularly consumer products. Batchelor (2001) gives an interesting view on how variety has developed in the automotive sector. MacCarthy, Brabazon and Bramham (2001) discuss how product architecture and operational strategies can impact and limit the potential customizability of a product.

2.5 Summary of the market environment

This wider perspective on the market environment has implications for how customization should be viewed in an MC context. Two points are stressed:

- Mass Customization does not automatically imply pure customization. Some compromise, limitations and constraints are inevitable if mass characteristics - responsive, efficient, high quality - are to be achieved and if premium prices are to be avoided.

- Customizable attributes evolve and change over time.

These considerations challenge the view not just of customization but how we view a product. We need to see operational systems generally and manufacturing systems in particular as delivering a service. In this wider context a value chain perspective provides key insights on Mass Customization.

3 Key value attributes and value chain perspective

3.1 MC value chain perspective

A production process may be viewed as a value chain or stream where every conversion process should 'add value'. Traditional value chain analysis forces a critical analysis of how and where value is added to the *product*. The MC value chain approach must take account of added value to the *customer* (MacCarthy 2001). The value chain perspective is premised on the view that customers aim to maximise value. Non-value adding activities generate only costs. It is important to note that variety in itself does not provide customer value.

Fundamental to a MC value chain perspective is the identification of the product attribute(s) that represent the *greatest perceived value* (real or otherwise) to the customer. These will be referred to as *Key Value Attributes (KVAs)*. They provide the basis on which to consider both customization potential and desirability. An attribute is a potentially customizable dimension of a product. The classification of attribute types is discussed further in section 5.1.3. One common example of a KVA is product colour, for example, for clothing and cars. Product colour is also an attribute that often poses significant challenges to operationalise.

Whilst the main focus for the conceptual KVA models are product attributes, service elements must not be neglected. For many products such as cars and photocopiers, elements of the service and after sales package could be considered to be KVAs.

3.2 KVAs as a focus for variety management

KVA analysis helps identify where the locus of variety should be. It follows that variety emanating from KVAs is relevant variety. Other product variation may represent non-value-added variety.

For any product attribute the following should be considered:

- Does the customer differentiate with respect to that attribute and how does the customer differentiate?

- Can the KVA be differentiated operationally?

- Is it desirable from the producer's and/or customer's perspective to differentiate on that attribute?

3.3   Constantly evolving KVAs

It is important to recognise that KVAs are not static because of the influence of product evolution. As discussed above, a range of factors will affect the stability of the customizable attributes – product maturity, fashion, markets, competition and technology. Product evolution and KVAs raise many fundamental research questions of importance to MC. At their heart is what the customer values and the level of demand for customized attributes. The demand may be latent or may change over time. It may not be absolute - there may be trade-offs e.g. how long a car customer is prepared to wait for precisely the right colour against whether they are prepared to pay a premium to have it delivered quickly. A current, potentially customizable attribute may represent a possible competitive advantage for a producer if it is achievable quickly or it may present an operational nightmare.

With respect to any product it is worth considering whether the KVAs have changed because of competitors and new entrants to the market, major technological changes, obsolescence, fashion, taste, demographics etc. Failure to analyse what customers value runs the risk of developing products with unnecessary variety or customization potential. Models and methods are needed to identify and analyse customization potential and desirability. Here we present a number conceptual models that help in this regard.

3.4   Conceptual customer KVA models

Figure 1 illustrates a conceptual approach based on differences in customer reactions to a range of product configurations. A product configuration is generated by a unique combination of product attributes. The total set of product configurations is represented by all possible permutations of product attributes. The perceived value of different configurations may vary across the customer base. The customer value of a product configuration reflects the customer's reaction to the product in its entirety i.e. the value of a combination of attributes. Although product configurations are discrete, in reality there are such a high number of product variants (e.g. personal computer configurations) that customer value can be represented by a curve. It is important to note that the value curves are only a snap shot of customer value and may not reflect the continually evolving perceptions of product attributes over time.

The potential for customization comes from differences between customers in how they value product attributes, illustrated conceptually in Figure 2. By comparing customers, a Value Difference curve (δV) can be considered, the shape of which may reveal whether and where customization is worthwhile, illustrated in Figure 2. Three different scenarios are shown below. A low level of value difference across configurations points to the suitability of a standard product with generalised functions and features. High levels of value difference would suggest that there is low homogenity in the market, thereby providing high customization potential. This customization potential may be difficult to exploit because of the diversity of values. Hot spots of value difference implies that there is a divergence in value perceptions for some product configurations. These concentrations of difference will give an indication of customization dimensions.

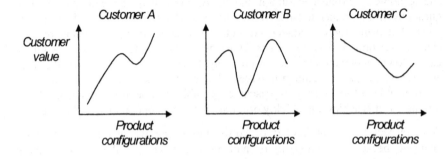

Figure 1: Conceptual illustration of customer differences

All Customers

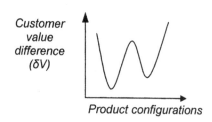

Customer
value
difference
(δV)

Product configurations

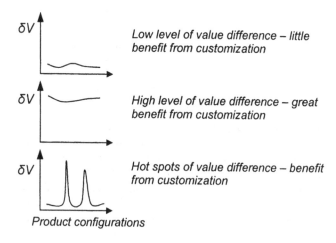

*Low level of value difference – little
benefit from customization*

*High level of value difference – great
benefit from customization*

*Hot spots of value difference – benefit
from customization*

Product configurations

Figure 2: The conceptual value difference curve with interpretations

The model in figure 3 illustates the problems with disrupting the continuum of product configurations. The use of a limited number of platforms may impose constraints on the feasibility of product configurations. For example, option x is only allowed with platform A and not platform B, but the customer values the product more highly with a platform B / option x combination. The impact of this will be that the customer will not obtain the greatest achievable satisfaction because customization is not available on the dimensions that are valued by them. Consequently there will be a loss in customization potential. Huffman and Kahn (1998) discuss related issues on how the customer perceives variety and how variety is presented to the customer.

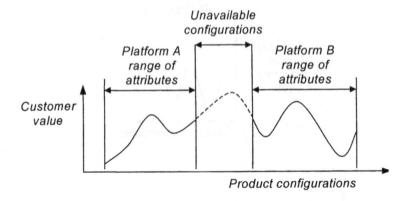

Figure 3: Inefficient variety – gap between two platforms with customizable attributes

3.5   Conceptual producer KVA model

But what is the optimum level of customization the producer should offer? Figure 4 illustrates a conceptual model to consider this problem. If customizable attributes are ranked and sequenced such that the most important across the customer base come first then we would expect the utility to the customer to increase rapidly (the solid curve) as the number of customizable attributes increases but, as the attributes become less critical, the incremental increase in perceived value reduces and a law of diminishing returns kicks in. From the producers perspective, the model assumes that the provision of a low level of customizable attributes may be achieved at a low level of additional cost but that costs increase at a faster rate as the number of customizable attributes increases because of variety related overheads (the dashed curve). The profileration in cost can be explained by the increase in complexity of the production system. A purely economic view then suggests that the producer needs to maximise the 'distance' between these two curves and offer customization around that level, suggesting the application of a Pareto-type analysis for customizable attributes. However Mass Customizing enterprises that are premised on delivering real customer value (at higher margins) are likely to operate in the higher region.

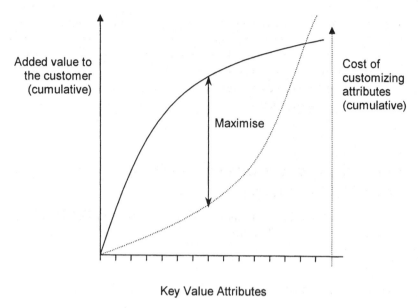

Added value to the customer (cumulative)

Cost of customizing attributes (cumulative)

Maximise

Key Value Attributes

Figure 4: A producer's model for customizable attributes

3.6 Putting the models into practice

The practicalities of constructing the various curves shown above have to be overcome. Eliciting customer ratings of value and estimating customization costs are non-trivial tasks. Once quantified, the value and cost curves may not be as well behaved and well defined as in the conceptual models, to the point where they are neither convex nor assymptotic.

It is a truism to say a product is more than the sum of its parts. Interactions between attributes may make a straightforward ranking of attributes impractical. It is feasible that a customer rates a certain product configuration as best value for money even with no single attribute matching their best option. This implies only a fraction of the total value of a product may be accounted for by individual attributes, in which case the cumulative curve in Figure 4 may a) not be representing the full value to the customer and b) understating the importance of a specific attribute by not accounting for how it adds value through interactions with other attributes.

Furthermore, offering customization on an attribute may make a product less good 'value-for-money' in the eyes of some customers. Customers who have no interest in customizing a particular attribute will be incurring a surcharge unless the manufacturer is able to customize that attribute without significant cost, delay or quality implications.

While there are hurdles to be overcome in operationalising the models above, they show a route to obtaining valuable insights. Customers can be daunted and put-off by too much variety (Huffman & Kahn 1998) which suggests a deliberate limitation of customizable attributes is warranted. For this reason the coherent framework of Value Difference and the cumulative value / cumulative cost ratio is of importance to mass customizers.

4 Operational and strategic implications of the value chain and KVA concepts

The value chain and the KVA concept provide useful perspectives for explaining how MC can be interpreted. They also shed light on operational practicalities and strategic choices of MC.

4.1 Operations implications

What process technologies, systems and organisational elements are required to implement the value chain? What competencies are needed for customizing a KVA, remembering that process competence may be as important as design or engineering competence. How should the supply chain be structured - suppliers, network alliances and logistics partners - in order to deliver appropriate levels of variety with respect to KVAs?

Identified KVAs influence the degree to which the customer is involved, and how they contribute to the product design and specification process. KVAs will also influence how and where customization is carried out. The point at which KVA customization is commenced may be a decoupling point in the operational system. Upstream of the decoupling point production may be forecast but downstream some degree of product differentiation with pull characteristics is likely. Non-KVA activities can be designed to be 'mass'. The challenge is to customize around the KVA in a 'mass' manner – fast, high variety, high quality, matched to customer requirements. Processes can be 'lean and mean' away from KVA points but this is not necessarily the case for the processes that deliver the KVAs in an MC system. They need to be flexible, responsive and capable of high levels of 'matched variation'. This may mean that low utilisation is inevitable for some resources. The real operational challenges arise when customization is required on multiple KVAs and where the variety overheads are most apparent.

A narrow interpretation of MC may assume simplistically that product allocation to customer must occur with the required customization specified very early in the operational system and that the customization component must be achieved within the factory. Together these imply highly inefficient and costly factories. From a value stream and KVA perspective, however, it should be clear that the customization component can be achieved anywhere in the extended enterprise. Thus customization can be achieved through suppliers, in specialised units such as micro-factories and in distribution channels. Indeed networks of focused factories can be used to deliver customized products. It shouldn't matter to the customer from where in the extended enterprise customization is delivered nor indeed whether the product was produced specifically for them – a stock item that meets their requirements exactly may be just as valuable as one that bears their name from order release.

4.1.1   Implications of different types of KVAs to operations

Different categories of KVAs can be distinguished and each type has implications for product design, process design, and inventory. There are five dominant types of customizable attributes, four of them to do with the product – *fit, function, aesthetic,* and *quality grade* – with the fifth being concerned with 'packaging'.

- Customizing the *fit* of a product often involves, but not always, adjusting a product's dimensions to the customer's specification or preferences. Clothing, fitted furniture and eyeglass lenses illustrate dimensional fit. An example of a different kind of 'fit' is matching food additives and dietary supplements to a customers needs. Dimensional fit requires a flexible process that supports quick set-up changes, whereas fit that is linked to the product's recipe has inventory and process knowledge implications.

- *Function* customization allows the addition of features or their removal from a product. This is enabled – or hindered – by a product's architecture. Using the taxonomy of Ulrich and Tung (1991), 'bus modularity' is preferable as it permits components (or assemblies) to be attached as and when necessary. Function customization has implications also for inventory (e.g. safety stocks) and for process management (e.g. equipment utilisation, line balancing with different cycle times).

- Colour is a common customizable *aesthetic* attribute and calls for process flexibility. Other aesthetic attributes such as drape in textiles and clothing, and styling attributes in consumer products may have many hidden operational costs, such as additional subjective human quality control processes.

- *Quality grade* customization gives customers discretion over the quality of the components, e.g. the speed of processor in their computer, or the choice of chrome plated versus stainless steel components in their kitchen. This has implications for product architecture to allow part interchange (categorised as component swapping modularity by Ulrich and Tung (1991) and for inventory management.

- Examples of customizing *packaging* include: using customer specified boxes and accompanying manuals; inserting additional promotional material and 'freebies'; supporting a two-for-one sales campaign. This attribute category has process and inventory implications. These require either flexible and re-configurable assembly systems, but more often means employing people to cope with frequent packaging changes.

Once the key value attributes are identified and classified, operational demands and risks can be assessed. Modular product design is an example of a demand that also carries risks. Achieving modularity requires more investment than might otherwise be called for (Baldwin and Clarke 1997) and is susceptible to being undermined by product evolution or revolution. For example, having invested in a design that permits customers to select functions, it may be galling to find that a technological breakthrough knocks down the cost to a level that results in a feature becoming a standard. The investment in designing a product that works with or without the feature and the development of a specification for an interface and housing may be lost.

4.2 Strategic issues

In a world with proliferating variety and brands, KVAs can be used in the analysis of customization competitiveness i.e. distinguishing useful variety from useless variety. It may also point to where variety will be tolerated - useless variety will be confusing and annoying whilst variety in KVAs will be desired and valued, probably better understood and likely to command higher prices. An important question for many organisations is who best understands the KVAs – designers, marketing professionals, sales, retailers, customer or the board of directors?

Value chains and KVAs provide a useful perspective to explain the nature of variety in particular sectors and typical new product development/new product introduction (NPD/NPI) strategies. They also provide a focus for operations strategy e.g. can company X compete by customizing about a specific KVA?

There is also the question of whether a company should offer customization of a KVA? Whilst operational limitations can place unwanted constraints on customization, strategic considerations can prompt voluntary restrictions. Product attributes link to a brand's image (Vriens and Hofstede 2000) e.g. a tough casing links to the 'safe and secure' value with which a brand has been aligned. Consequently, the extent to which the casing can be customized should be controlled. It may be feasible operationally to offer various casing materials, but would it undermine the brand strategy to do so? For some attributes it will be a yes / no answer, but for others the strategy will allow a degree of customization. The problem becomes one of defining an acceptable envelope of KVA customization.

5 Conclusions

An examination of the MC market environment exposes a range of issues that are relevant not only within the MC domain but poses questions for those following different strategies such as response to variety proliferation and reaction to product evolution.

Looking upon Mass Customization as just a fusion of *mass* plus *customization* is too narrow a view – mass customization is not pure customization for everyone (MC ≠ M + C). Mass customization entails some compromise if customization is to be delivered on a mass scale. However we do need to clarify the spectrum of operational and business strategies that should come under the MC banner. Differentiation and customization are achieved through Key Value Attributes that reflect real customer value.

The benefits of the KVA concept are:

- KVA allows judgment of whether customization is relevant.

- Knowing a product's KVAs focuses attention on worthwhile customization and avoids effort being channeled toward unmerited variety.

- KVAs can be classified into different types of attributes to give indications of the types of product architecture and the types of order fulfillment processes and their control required for customization.

It is recognized that there are difficulties in applying the KVA concept. Extracting KVAs should not be understated due to problems in evaluating customer value and disentangling attribute interrelationships. However, analysis of KVAs through value difference provides a useful perspective on the customer base according to customizable attribute potential.

A value chain perspective of the supply chain is needed to understand where customization can take place, how it can take place and its likely impact. MC may be delivered by the extended enterprise - suppliers, networked alliances, production networks or logistics partners. Essentially it does not matter where or how or even when the requisite customised value is added, provided it is delivered. The acquisition of true MC capabilities by any enterprise in a sector has the potential to radically change the competitive landscape.

Acknowledgements

We would like to acknowledge our project research colleagues at Oxford University – Claudia Mchunu, Janet Efstathiou and Aruna de Alwis. We are very grateful for the support provided by Engineering and Physical Sciences Research

Council (EPSRC), Systems Integration Initiative (project GR/N11742/01) and, not least the support provided by our industrial consortium.

References

Agrawal, M. Kumaresh, T. Mercer, G. (2001): The false promise of mass customization. The McKinsey Quarterly 3:62-71

Baldwin, C.Y., Clark, K.B. (1997): Managing in the age of modularity. Harvard Business Review 75:84-93

Batchelor, J. (2001): Engineering a vehicle for world class logistics. In: MacCarthy B & Wilson JR (eds) Human Performance in Planning & Scheduling, Taylor & Francis, London, pp 355-385

Coates, D. (1998): Achieving integrated order management in a major steel business. Paper Presented at IEE Responsiveness in manufacturing, London

Cox, W., Alm, R. (1998): The right stuff: America's move to mass customization. Dallas

Fisher, M.L., Ittner, C.D. (1999): The impact of product variety on automobile assembly operations: empirical evidence and simulation analysis. Management Science 45:771-787

Hayes, R.H., Wheelwright, S.C. (1984): Restoring our competitive edge: Competing through manufacturing. John Wiley and Sons, London

Holweg, M., Greenwood, A. (2000): Product variety, life-cycles and rate of innovation - trends in the UK automotive industry. Paper Presented at Logistics Research Network, Cardiff, UK

Huffman, C., Kahn, B.E. (1998): Variety for sale: mass customization or mass confusion? Paper Presented at Marketing Science Institute Report, Cambridge, MA

Kekre, S., Srinivasan, K. (1990): Broader product line: a necessity to achieve success? Management Science 36:1216-1231

MacCarthy, B. (2001): Multiple perspectives on mass customization: definitions, concepts and classifications. Keynote Paper Presented at 2001 World Congress on Mass Customisation and Personalisation, Hong Kong

MacCarthy, B., Brabazon, P.G., Bramham, J. (2001): Mass customisation: alternatives to using product modularity as a route to mass customisation. Paper Presented at 16th International Conference on Production Research, Prague, Czech Republic

MacDuffie, J.P., Sethuraman, K., Fisher, M.L. (1996): Product variety & manufacturing performance: evidence from the international automotive plant study. Management Science 42:350-369

McDermott, C.M., O'Connor, G.C. (1995): Managing in the age of the mass merchant. Business Horizons 38:64-70

Pilkington, A., Chong, D. (2001): Conflict at the interface: Mass customization definitions for operations, marketing and strategy. Paper Presented at Euroma, Bath, UK

Stalk, G., Jr. (1988): Time - the next source of competitive advantage. Harvard Business Review 66:41-51

The Economist (2001): Wave goodbye to the family car. 358:57-58

Ulrich, K., Tung, K. (1991): Fundamentals of product modularity. Paper Presented at ASME Winter Annual Meeting Symposium on Issues in Design/Manufacturing Integration, Atlanta

Vriens, M., Hofstede, F.T. (2000): Linking Attributes, Benefits, and Consumer Values. Market Research 12:4-10

Womack, J.P. (1993): Mass Customization: The New Frontier in Business Competition. Sloan Management Review 34:121-123

Yeh, K-H., Chu, C-H. (1991): Adaptive strategies for coping with product variety decisions. International journal of operations & production management 11:35-47

Zipkin, P. (2001): The limits of mass customization. Sloan Management Review 42:81-87

Part II

Preparation and Implementation

Web Tools for Supporting Mass Customization

Aldous, K. J. & Nicholls, H. R.

Summary: A brief overview of state of the art of mass customization is presented, followed by general architecture for mass customization of manufactured products. The tools required to realize the architecture, and the relationships among them are described. The requirements for document description standards both within the factory and among the business community are outlined, then each of the significant components and tools needed to support the architecture are briefly described. These tools include product configurators, tools to translate customer-configured products into the business objects required for manufacturing and plant management, customer authentication, supply chain management, and by-products of mass customization technology, such as the dynamic promotion of products. The role of XML and related technologies in enabling the development of these tools is discussed.

Keywords: Mass customization, Internet, World-wide-web

1 Introduction

Mass customization aims to provide goods and services that best meet individual customers' requirements with near mass production efficiency. It is a new technology whose definition is evolving as supporting technologies and capabilities are developed.

In this paper, we focus on systems in which communications between the customer and the enterprise, and among the several operational units of an enterprise are mediated by Internet-based software. This software allows both the customer to specify detailed product requirements, and the enterprise to generate and exchange properly constructed documents needed during the production of the specified products. Since each customer's requirements may be quite distinct, the challenge is to produce a tailored product that meets a unique specification. When the requirements are complete, they are translated into manufacturing requests that are scheduled for execution or purchase.

The manufacturing facility must track each job, relate it back to one or more orders, report on its progress, and ensure that it is completed on time to meet agreed delivery dates. Order fulfillment and accounting operations must be informed when orders are ready for shipping and the appropriate documents prepared and dispatched with the goods, electronically, and perhaps by mail. Communications along the supply chain must also be generated and transmitted, and replies received and interpreted. We discuss the use of emerging technologies for the manipulation and exchange of information among disparate manufacturing information systems, and describe a prototype systems that has been developed to demonstrate these technologies.

We briefly review the current status of mass customization technology, discuss its future evolution and describe the form of some of the tools that are needed for mass customization to become a commercially functional reality.

An outline of an architecture for a general mass customization system, and a description of tools that we have developed that will contribute towards the realization of this architecture are presented.

The concepts for these mass customization tools are based upon our experiences in developing a method and web-based tools for supporting groups of small-to-medium manufacturers who are collaborating in the definition and manufacture of products (Nicholls *et al*. 1999). The tools have been, or are being, tested in other contexts and, from initial results, are effective and robust.

2 The State of the Art

Mass customization is easier for some products than others. The personal computer is an example of a product well-suited to customization, and Dell computer at their web site http://www.dell.com/ us/en/gen have successfully done this; customization of shoes is more difficult, as Nike discovered (Wilson 1999).

Many currently deployed applications of mass customization have been developed from electronic commerce front-ends, and are often confined to product specification using intelligent catalogues (CAP/Sweet's Group 1999) and compiling customer relationship management databases for fine-grained market segmentation.

Mass customization is more than these two things, valuable though they may be; it is also more than the customized assembly of modular components, as proposed by (Duray 1997). Indeed, it is in the manufacture of products other than those assembled from cut-to-fit or standard modules that mass customization presents its greatest challenges, such producing customized lots of woven fabrics, as discussed by (Sanders 1999). Nevertheless, implementing Duray's more limited definition is a demanding task as shown by the organizational and technical difficulties of customizing footwear as described by (Miller 1996) and more recently by (Wilson 1999).

Drivers for mass-customization include the expectations generated by the immediacy that the Internet provides. More or less instant access to information and communication drives a demand for swift access to tailored consumer products, and the Internet, in turn, provides the means by which that demand can be met.

3 An Architecture for Mass Customization

The critical tools needed for a mass customization system include:

- one or more product configurators, composed of server and/or client side web pages and code, that enable the customer to specify what they want;

- a customer orders database;

- a translation system that analyses customer orders, generates shop orders (that is, manufacturing tasks that can be scheduled), updates the inventory, issues purchase orders and computes expected order fulfillment costs and dates for customers;

- a system for customer authentication;

- a customer relationship management database;

- a system to support inter-company communication for supply chain management;

- a reconciliation system for order fulfillment and billing.

The tools and databases that are needed for these tasks and the relationships among them are shown schematically in Figure 1. The heavily outlined boxes are the principal tools. They include web-based product configurators, customer authenticator, supply chain manager, and the three components of the translation system for product analysis, shop and purchase order generation and order fulfilment prediction for informing the customer of delivery times and costs.

Also shown in Figure 1 is a tool for dynamic product promotion. The mass customization system needs to have access to inventory and other supply information in order to predict fulfilment. This information can also be used to generate real time promotions such as offering discounts for products that, for operational reasons, are convenient to produce or are simply in excess supply.

This architecture does not specify the connection between the shop order databases and the shop floor, where further technology to enhance the manufacturing processes may be required.

In concentrating upon the delivery of customized goods and services, the architecture does not address the issues of customer support throughout the use of the product – mass customization includes post sales support, and information to assist the customer enjoy the use of the product. Information gathered during the sales and fulfillment processes is stored in the customer relationship management database. This provides a resource for marketing and customer services, and should be used to personalize responses during subsequent visits to the web site by the customer.

4 Document Description Requirements

The Internet is the principal enabling technology that allows the customer to communicate with the manufacturing system. Recent and continuing innovations in the functionality of browsers, especially Microsoft Corporation's ActiveX technologies, dynamic HTML, the Document Object Model (DOM) and Java enable the complex information processing operations needed for product specification to be performed using local resources, thus reducing the dependence upon the responsiveness of the server and the communications medium.

The main innovations of the architecture shown in Figure 1 are the use of this dynamic facility in downloaded pages combined with local data resources, and the translation system. To be generally applicable, the translation system requires standard Extensible Markup Language (XML) definitions for document exchange throughout the manufacturing facility. For a given application, *ad hoc* forms may be used, but general solutions, based on XML, need to be sought, as pointed out by (Bosak & Bray 1999).

In addition, the supply chain requires standard XML definitions for document formats and content for documents such as orders, invoices, inquiries and other business objects. These issues are being addressed, although the diverse initiatives such as BizTalk (http://www.biztalk.org), ebXML (http://www.ebxml.org), PIPs (http://www. rosettanet.org) and XEDI (www.xedi.org) will perhaps delay the development of standards by confusing the situation with too many alternatives.

Equally important as standard definitions of document formats and content is the ability to select information from a given document and convert it into the forms required by other processes. We discuss some emergent technologies for performing translations of this kind in the next section.

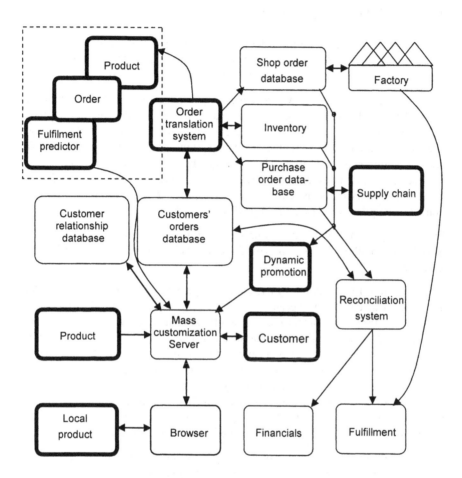

Figure 1: Tools for a mass customization system

5 Tools

In this section, we discuss each of the components represented by the heavily outlined boxes in Figure 1, and describe how XML-based technologies may be used to perform the data extractions and document transformations needed to generate the documents of the form required by the various processes.

5.1   Product Configurators

Internet-based tools that we have developed for document exchange and product specification (Aldous & Nicholls 1999 and Nicholls *et al.* 1999), are examples of the technologies required for Internet-based product configurators.

Among these tools are some which consist of web pages with embedded code that manipulates the DHTML object that is displayed by Microsoft's Internet Explorer. This code is executable by the user *via* mouse or keyboard functions. An example is a page containing an extendible quality function deployment (QFD) matrix for product requirements definition. This page may be stored locally and used over an extended period to develop a set of requirements and features that are stored in the page. Once the user is satisfied with these, the page, containing the user's definition may be uploaded to a server for processing.

Extendibility of the QFD matrix is implemented using the programmable features of DHTML to allow the user to expand the matrix, expressed as an HTML table, as required.

The dynamic HTML technology used in this page may be adapted for any product at any level of detail, and because such pages are initially accessed from a web server, their form and content may be rapidly and easily changed to suit product variations, stock availability, the state of the schedule and so on. A major advantage of this method is speed of response to the user's actions. By encoding as much of the configuration system in the transmitted web page as is feasible, and perhaps storing some of the components permanently on the client's site, communication traffic is reduced, and response therefore improved. More information needs to be downloaded at the beginning of the process, and a trade-off between the size of this initial download and repeated site accesses must be made.

One of the important attributes of a configurator, as pointed out by (Duffel and Street 1999) is the ability to capture knowledge efficiently. A mass customization configurator must include means rapidly to capture and amend product models and configuration rules, and of generating web pages containing embedded code that present the rules to the user. Commercial product configurators that include facilities for the first of these tasks are available, but systems capable of generating efficient, web-enabled configurators are yet to appear.

5.2 Order Translation

Once a customer has configured the desired product, and has had an order provisionally accepted, the order must be translated into a bill of materials and a set of shop orders. The inventory database also needs to be queried and purchase orders issued if necessary. The latter may require querying other members of the supply chain, and methods for doing this in real time over the web are among the critical technologies that need to be developed into easily deployed components.

The product model presented to the customer for configuration will contain a simplified subset of the data required for manufacture, and it is the task of the configurator to validate the customer's input and generate a product data model that includes a bill of materials and routing.

Validation must ensure that the received order is correct and that the customer has not forgotten anything important. At this point, a price for the specified product will also be generated.

Finally, order fulfillment prediction based on the current schedule and using statistical demand prediction techniques must provide an expected delivery date for the customer. Once the customer has been presented with a price and a delivery date, he or she must either confirm the order, or cancel the process.

5.3 Customer Authentication and Payment

Methods for customer identification and authenticating the financial information supplied during purchasing procedures, such as log on for registered customers and credit verification using information in the customer relationship database, are well understood. Similarly, order acceptance and filing processes, and funds transfer using secure payment technology are currently in use for electronic business operations, and can plainly be employed unchanged for mass customization systems.

5.4 Supply Chain Management

Another of the tools that we have developed is a web-based system to enable geographically dispersed groups to exchange and manage documents generated during joint projects. In this system, documents associated with a project or other undertaking are stored at web site, and each has a designated owner. Documents may be accessed only by specified registered participants in the undertaking. Access to any particular document is also controlled by the owner, and a trace of all accesses to each document is maintained. The principles of this tool could be used to provide a platform for document interchange required for supply chain management.

In addition, to enable communications among supply chain members to be effective in real time, as required, for example, for accurate order fulfilment prediction, a method for the automatic access of remote applications is needed. There have been a number of initiatives to fulfil these functions, including commercial products, and proposed general methods such as Web Interface Definition Language (Merrick and Allen 1997). It is clear, however, that products based on a protocol such as Simple Object Access Protocol (Box *et al.* 2000) will provide the open, universal access mechanism needed for general business to business communications. This technology, and the standard XML definitions for business objects discussed in section 4 will together provide the infrastructure required for real-time supply chain management.

5.5   Technology for Document Transformation

As we have mentioned, document exchange throughout the manufacturing facility requires both standardized document formats and content, together with capabilities to select and convert the information in a given document into other forms, and to transmit the converted documents to other applications for processing.

A prototype system has been developed (Aldous *et al.* 2001) that demonstrates how the exchange of documents between disparate applications in a distributed manufacturing environment may be effected. The prototype uses two key technologies: Java servlets (Davidson and Coward 1999) and Extensible Stylesheet Language (XSL) transformation (Clark 1999). Java servlets are software modules that perform server side processing of HTTP requests, typically from a WWW client. For example, a servlet may process information gathered from a Web browser and transmit a new web page back to the browser with updated information.

XSL, on which XSL transformation technology is based, serves two purposes: to format XML documents for rendering, and to transform XML documents into new forms. These two purposes are closely related. XSL transformations are intended, *inter alia*, to be used to select and order elements of XML files as a preliminary step to formatting them, as specified by a XSL document (which is itself a special kind of XML document), for rendering in a web browser. This selection and ordering process may, however, be used for more general document processing and exchange purposes, such as the transmission of messages among the components of a manufacturing information system.

Transmitting information from one information processing application to a second, different application involves three steps:

- generating an XML document from an arbitrarily formatted report produced by the first application;

- selecting and ordering the content of the XML document according to the requirements of the second application;

- generating a document suitable for processing by the second application.

The report generated by the first application may, of course, be prepared as an XML document, thus eliminating the first step, but where this is not feasible, products and methods are available for converting arbitrarily structured character files into XML form. *XML Convert* (Unidex Inc. 2001) is an example of a commercial product available for this purpose. The third step may generate a document that, apart from a minimal set of compulsory XML tags, is largely in the form required by the application, and only requires these tags to be stripped prior to final processing (or simply ignored by the application).

In ASCII character form, XML documents cannot be easily manipulated by computer programs. Since they represent a structured hierarchy of nodes, however, they may be composed into a data structure that can be processed programmatically. The Document Object Model (W3C DOM Working Group 2001) has been defined for this purpose. XML documents in this form are called Document Object Model objects. Application programming interfaces (APIs) have been written to construct DOM objects from character streams, such as files or strings containing XML documents, to manipulate these objects and to convert them back to character streams. An example is the Java API (Mordani *et al.* 2001).

Transformation of DOM objects may be performed by a DOM API using the content of one or more XSL documents to direct the process. The rules for transforming a document for exchanging information between two applications in a manufacturing system are thus encoded in an XSL document, and require no programming code. A single XML document, generated from some application, may be used as the source for several quite different destination documents required by other applications simply by specifying an appropriate XSL document for each destination document.

5.6   Prototype Document Exchange System

The facilities provided by the Java API for XML processing were used to construct the prototype document transformation and exchange system described in (Aldous *et al.* 2001). This system comprises a group of software modules that may be deployed on one machine, or be distributed on an intranet or the Internet.

The central component of the system is called the aggregator module. This module acts as a junction for information transfers between other system components. It communicates with a series of modules called proxies, each of which serves as an interface to an information resource, such as a remote software module, a proprietary package using a non-standard data format or another organization's information system. A proxy may perform some or all of its processing itself and therefore act as an application in its own right. The aggregator is only accessible through a proxy, and it treats users no differently from any other information resource.

Messages transmitted to the aggregator by a proxy are constructed from information provided by a user or other information source using an XSL transformation, as are documents that need to be rendered in a browser. The URIs of the XSL documents used for these transformations may be included in the XML messages themselves.

Messages received by the aggregator will include segments destined for one or more proxies. The aggregator scans messages for tags indicating this, and redirects each segment to the appropriate proxy as determined by a uniform resource identifier (URI) contained in the message segment. The proxy in turn will process this message and direct the result to the application for which it acts, compose any response from the application into a message and transmit this back to the aggregator. The aggregator maintains the state of an extended transaction, and will continue or complete the dialogue with the initiating application or user.

5.7   By-Products of Mass Customization

With any new technology, additional benefits often occur as by-products. Sometimes these do not become evident until the technology is used. However, two that may arise from mass customization systems are the use of information that is accumulated by the product configurator for product adaptation, and the dynamic promotion of product features and discounts using real-time information from the shop order and inventory databases and from the scheduling system. For example, where certain operations that require lengthy and expensive set-up are already scheduled, it may be possible to produce product or product features that require the set-up at a significantly reduced marginal cost, and these products or features may consequently be offered as special deals.

6   Conclusions and Further Work

Mass-customization is a technology that exploits the latest advances in communications and computing technologies. The architecture and Internet tools that we have described offer support for realizing the model for mass customization presented in this paper. The ultimate goal of supplying what the customer requires, and nothing else, will be met with increasing accuracy as the technology is developed.

We are currently undertaking a project, in collaboration with a large company, to validate the architecture and tools discussed in this paper, and to generate a plan for their implementation. We have selected some mass customization scenarios relevant for the products the firm produces and the market segments that it supplies, and we are currently constructing detailed use cases to capture the requirements of a system to realize these scenarios.

Acknowledgement

This work was undertaken as part of contract C0X8009 with the New Zealand Foundation for Research, Science and Technology. The authors are indebted to our colleagues Alan Caughley and Andrew Lintott, who devised and constructed the prototype document exchange system.

References

Aldous, K.J., Lintott, A. and Caughley, A. (2001): Electronic Business for Manufacturing. Industrial Research Limited report 1039; Christchurch, New Zealand

Aldous, K.J. and Nicholls, H.R. (1999): An Internet tool for supporting inter-company collaboration.. Proc. 26th International Conference on Computers and Industrial Engineering, Volume 1, pp 229-234. Melbourne, Australia. Dec. 15-17

Bosak, J. and Bray, T. (1999): XML and the second-generation Web. Scientific American, May 1999: 79-83

Box, D., Ehnebuske, D., Kaklvaya, G., Layman, A., Mendelsohn, N., Nielsen, H.F., Thatte, S. and Winer, D. (2000): Simple Object Access Protocol (SOAP) 1.1. Retrieved Nov. 8, 2000 from http://www.w3.org/TR/SOAP

CAP/Sweet's Group (1999): CAP Products Offices Online. Retrieved Nov. 8, 2000 from http://cap.sweets.com/ prodoo.htm

Clark, J. (1999) : XSL Transformations (XSLT) Version 1.0. Retrieved July 11, 2001 from http://www.w3.org/TR/- xslt

Davidson, J.D. and Coward, D. (1999): Java™ Servlet Specification, v2.2. Sun Microsystems Inc, Palo Alto, CA

Dodds, L. (2000): XML reduced 1. Retrieved Nov. 8, 2000 from http://www.xml.com/pub/2000/10/11/deviant/index.html ?wwwrrr_20001011.txt

Duffell, J. and Street, S. (1999): Mass customisation across the business: configurators and the Internet, Part 2. Control, Dec. 1998/Jan. 1999: 14-16

Duray, R. (1997): Mass Customization Configurations: An Empirical Investigation of Manufacturing Practices of Customization. Ph.D. Dissertation, Ohio State University, Ohio

Merrick, P. and Allen, C. (1997): Web Interface Definition Language (WIDL). Retrieved July 12, 2001 from http://www.w3.org/TR/NOTE-widl

Miller, M.J. (1996): If the shoe fits... PC Magazine, **15**(18)

Mordani, R., Davidson, J.D. and Boag, S. (2001): Java API for XML processing. Retrieved July 12, 2001 from http://java.sun.com/xml/jaxp-1_1-spec.pdf. Sun Microsystems, Inc. Palo Alto, CA

Nicholls, H.R., Aldous, K.J. and Cook, C. (1999): An information and technology management method for groups of small engineering companies. Proc. 26th International Conference on Computers and Industrial Engineering, Volume 1, pp 129-135. Melbourne, Australia. Dec 15-17

Sanders, F.-H. (1999): How to cope with ever smaller lot sizes in weaving. International Textile Bulletin, 3/99

Unidex, Inc (2001): Overview of XML Convert & Xflat. Retrieved July 12, 2001 from http:// www.unidex.com/overview.htm

W3C DOM Working Group (2001): Document Object Model (DOM). Retrieved July 12, 2001 from http://www.w3.org/DOM/

Wilson, T. (1999): Nike Model Shows Web's Limitations Retrieved Nov. 8, 2000 from http:// www.internetwk.com/story/ INW19991203S0004

Manufacturing Planning and Control Content Management in Virtual Enterprises Pursuing Mass Customization

Rautenstrauch, C., Tangermann, H. & Turowski, K.

Summary: While pursuing mass customization (MC) strategies, companies often benefit from cooperating with others. From a customer's viewpoint, cooperating companies become a virtual enterprise (VE) that produces customer-individual goods. Participants of such a VE have to be extremely flexible in planning and executing their business processes. They have to be coordinated and supported by information systems (IS) that are particularly designed for these tasks. By focusing on the MC supply chain the basic data that is relevant to make successful MC strategies possible is identified. At first, the MC-macro-process, the supporting IS, and the change of roles of the participants during this process will be discussed. Based on this role concept and the common model of a VE for MC, relevant enterprise resource planning (ERP) data is identified.

Keywords: *Mass customization; enterprise resource planning; virtual enterprise; coordination of production*

1 Mass Customization and Virtual Enterprises

By *mass customization* (MC) is meant a synthesis of mass production and the manufacture of customized products [24, p. 48]. If firms previously faced the dichotomous choice of whether to produce cheap mass products or to become custom manufacturers producing customized products at relatively high cost, MC aims at combining the positive characteristics of both alternatives: the (customized) production of individual goods at the price rates of mass production with the highest product quality and shortest delivery times. From the customer's viewpoint, in the ideal case products manufactured using MC differ neither in price nor quality nor delivery time from ones produced by mass production (MP). The sole difference should be the precise fulfillment of individual requirements.

However, in practice it is hard to define the exact borderline between MC and MP with a lot of possible variants or single-item production. One way to define this borderline between MC and single-item production is to determine a threshold value for cost per unit of output [22, p. 65]. If the cost per unit of output for an individual pro-duct is not higher than approximately 10-15 percent of a comparable mass product, we speak of MC, otherwise of single-item production. Compared to single-item products a bill of materials of a MC product contains a higher portion of standard parts and assemblies. Furthermore, in MC parts of the final products are individually designed (not only configured!) based on customer's demands.

Given worldwide slow market growth, particularly small and medium enterprises (SME) are increasingly in the unattractive position of having to face predatory competition from large enterprises which employ MC in markets originally reserved to them. While the situation in large enterprises is first characterized by the necessity of increasing the flexibility of previously relatively rigid production processes, SME must primarily structure often already project-oriented manufacturing more efficiently. For this purpose, they need to reorganize their production and expand their use of information technological support in the production domain.

Due to the economies of scale principle, large enterprises retain the advantage of large size, even if SME are technically and organizationally optimally adapted to MC. Therefore, especially for SME it is necessary to enter into alliances with other SME in order to acquire the necessary counterweight to hold their own in competition. In particular, suppliers should be integrated into a dynamic network with an organic organizational structure, but as well it should include enterprises like construction firms or software developers which could support the product development and manufacturing process.

In order to provide the necessary flexibility in such alliances the partnership should be as loose as possible, i. e., it should be formed for a limited period of time and a limited product palette. From the viewpoint of the customer who buys such a product, the alliance seems to be a single company. In fact, it is a *virtual enterprise* (VE). The concept of VE has its origin in the idea of the *agile enterprise* [19, pp. 8-9]. It is a temporary network of independent companies or natural persons, which have the same rights, and that cooperatively produce a specific good or service. The collaborators particularly bring in their core competencies, and there is no institutionalized common management. The VE appears to others as a single company, and it is coordinated through the use of a suitable IS, cf. e.g. [1, p. 10]. However, it is possible that over time the VE becomes a conventional group or fusion of companies, if it is successful with its mass customized products in the market for a longer period. In this case, the period in which a VE exists is a preparation phase for establishing a steady cooperation of companies. In such cooperations, the companies are connected through "normal" supplier-producer-relationships. The process of changing a VE to a steady cooperation can be incremental, if it starts with a core group of members of the VE which is extended by adding more members of the VE to it over time. This does not exclude the case that some companies stay loosely coupled to the cooperation if their products or skills are demanded sporadically only.

The founding, coordination and dissolution of such VE are likewise essential management tasks which are performed in mechanical organizations only in the exceptional case. For this reason MC basically requires an *organic* organization [23, p. 111].

The foundation of MC is provided by dynamic networks consisting of relatively autonomous teams (*task forces*) which interact with other teams independently of individual product specifications [23, pp. 115-119]. Characteristic of this type of organization is that the relationships between individual teams and their makeup are subject to continual change. Employees in such networks are not only integrated into project work, but also into operative planning and quality control. This requires a higher level of qualification than in the case of a mechanistic organization and readiness to assume responsibility in the completion of projects.

Participants of VE which are set up for MC of a particular product line have to be as flexible as task forces in a real company - up to building a VE on demand for each customer-individual product. From an outside view, a VE has to work like a conventional enterprise. In practice, a VE is a co-operation of different real world companies. They have to be coordinated and supported by IS which are especially designed for these tasks. In particular, the special aspects of ERP-systems are discussed in this paper. Therefore it is necessary to illustrate at first the organizational prerequisites, i. e. the MC-macro-process and the changing roles of the participants during this process.

The MC-macro-process is embedded into the framework given by a general model of a VE for MC (cf. figure 1). It consists of a MC-macro-process, IS support related to sub-processes, the changing roles of participants in the process, and a flexible and open coordination concept.

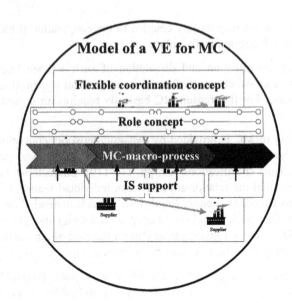

Figure 1: VE model for MC

2 The MC-Macro-Process

The process begins with the *configuration of a desired product* by the customer. After that, it is necessary to check which parts or assemblies can be manufactured by the producer and which must be *acquired*. For parts which must be acquired individual attributes may possibly also have to be taken into account. In such a case the enterprise assumes the role with regard to its supplier of a customer with individual requirements. The supplier can also assume the same role with regard to a further supplier.

Figure 2: MC-macro-process

This leads to a nesting of procurement activities and a networking of the production processes of suppliers and manufacturers. Procurement follows the *production* of individual products by flexible or respectively partly-automated groups. The data basis for this is provided by *ad hoc generated bills of materials and work schedules*, which implies that long- and mid-term overall planning in the usual sense is impractical for MC, since given the great uncertainties in regard to the products which must be produced in a longer planning period, realistic parameters cannot be determined. Production also includes quality assurance. The process is completed through the *physical distribution*.

Macro process

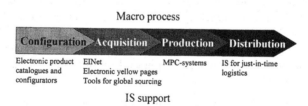

Electronic product	EINet	MPC-systems	IS for just-in-time
catalogues and	Electronic yellow pages		logistics
configurators	Tools for global sourcing		

IS support

Figure 3: IS support for MC

The following list sketches briefly which steps of the macro-process can be supported by which types of IS [17]:

- *Configuration of customized products*: Here *configuration tools*, e.g. WWW-based (WWW stands for World-Wide-Web) electronic product catalogues (EPC) and product configurators [16], can be employed based on electronic product catalogues. Such systems support the user to the extent that they display all available components as well as reasonable combinations of them and exclude incompatible configurations. Accordingly, consulting systems are often integrated into such configurators. While configuration tools particularly support customers and marketing employees, computer-assisted design (CAD) systems can be expanded with functions of *Custom Data Management* (CDM). On the basis of product configurations desired by customers it is determined in which areas an existing construction must be modified to customize it.

- *Procurement*: The coordination of procurement activities (Supply Chain Management [12]) up to and including the founding of VE can occur in various ways on worldwide networks, and here especially via the Internet. Promising approaches for the support of supra-enterprise MC activities [15] include the establishment of electronic "yellow pages" such as the Enterprise Integration Network (EINet), through which manufacturers and suppliers can locate one another, on up to worldwide activities of *global sourcing*, in which middlemen mediate between producers and suppliers worldwide via the Internet.

- *Production*: Manufacturing planning and control (MPC) systems (as part of ERP-systems) must be in the position to generate short-term ad hoc bills of materials and work schedules on the basis of which production projects can be scheduled in real time. Furthermore, such systems must be able to support the short-term selection and coordination of teams [26, p. 402].

- *Physical distribution*: Here *just-in-time concepts* are particularly to be supported by appropriate IS.

3 Participants and their Roles in a VE for MC

The first group of participants of a VE for MC are (small and medium) manufacturing enterprises and their suppliers, who employ MC, too. In case of MC, a significant portion of knowledge about design and quality of the final product is often owned by customers and merchants or brokers. Hence it follows that this group has to be included into the VE for MC, because they are needed for quality assurance in the production process. Figure 4 illustrates the change of roles during the MC-macro-process.

At the beginning of the process, the roles are distributed as follows:
- The *customer* (C) is the person or company, who wants to buy the final individual product.

- The *producer* (P) is the company, who offers this individual final product.

- The *merchant* or *broker* (M) supports customer, producer, and suppliers with know-how about design, configuration and quality characteristics of the individual final product.

- The *suppliers* (S) deliver parts or assemblies to the producer. Only those suppliers are relevant, who employ MC themselves.

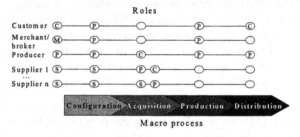

Figure 4: Changing roles in the MC-macro-process

Since customers and (possibly) merchants/brokers are involved in the product design and often also in quality management if MC is applied, they become producers in the configuration and production phases. Producers become customers in relation to their suppliers in the acquisition phase, if the suppliers apply MC. Since suppliers might also have other suppliers, their role might change to the customer role in relation to their suppliers. Figure 4 depicts such a relationship between supplier 1 and supplier n. In other words, with each link of a supply chain a new MC-macro-process is set up, and roles are changing.

Figure 4 shows that in principle all participants can play each role in a VE. Therefore, ERP-systems become of strategic importance in such environments, because all parts of the value chain are supported by them and it becomes necessary to discuss their adaption to VE for MC application.

4 Connecting ERP-Systems for MC

Most of traditional ERP systems handle production as completely internal to the company. Other business areas as materials management, marketing, sales and distribution, management and financing, and controlling are production's interface to other companies that are part of a VE for MC.

MC forces the production department of a company to communicate with business partners in order to optimize business processes. Figure 5 display possible communication partners of the production area. It gives an overview about possible communication partners of ERP-systems. At first, communication can be classified concerning an internal or external partner.

Data exchange to internal partners is mostly realized through the integration of the business areas into a business standard software. Intranet technologies can also be applied for this communication.

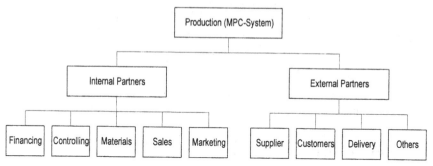

Figure 5: Possible communication partners

External partners receive information through messages or get read-only access through a specific interface. Data input into ERP-systems by external partners are rare – except partnerships, where one partner has enough market power to force the others to grant write permission. An example for these types of partnership is the car manufacturing industry. Suppliers have to allow the car manufacturer to place orders in their ERP-systems.

Production data can be transmitted in two different communication modes: active (*a*) or passive (*p*).

Providing production data is called *active* if the ERP-system is the data transfer initiator. For example, an electronic mail (e-mail) or electronic data interchange (EDI) messages are sent to the recipient containing the information in question [32].

If production data is provided using an on-request basis, the approach is called *passive*. Persons or systems request the needed ERP-data using an Internet application or sending an EDI message. The requested information is returned by the production system.

Table 1: Classification of production data exchange

Communication Partner	Information	a	p
Supplier	Material requests (JIT)	+	
	Planned changes of needed material	+	
	Product configuration information	+	+
	Planned material delivery		
Customer	Planned delivery date	+	+
	Product availability		+
	Product configuration information		+
Delivery	Planned pick-up time	+	+
	Delivery destinations	+	+
	Delivery content information	+	+
Others	Product information	+	+
	Capacity load information	+	+

Table 1 gives an overview about external communication partners and the appropriate communication issue. The table will be explained in the following sections.

4.1   Bill of Materials

This information is usually part of *product data management* (PDM). PDM-systems contain also geometry, design documents, and product structure (bill of materials) [5, p. 84].

E-commerce systems using a build-to-order approach manage often a vast number of possible product configurations. However, not all combinations are logical and producible according to physical, technical, production, and legal constrains.

Taking the example of an e-commerce site offering customized cars, it is not possible to build a convertible with a sliding roof. Product and variant management of ERP-systems have to identify this combination as not possible.

Bill of material information are also of importance in case of a alteration, introduction or cancellation of a product. Suppliers need to prepare to these upcoming changes. This friendly behavior can help to build a tight relationship between business partners and prevent the occurrence of delivery bottlenecks of the needed materials.

This functionality is provided by *electronic product catalogues* (EPC). EPCs support presentation, consulting, configuration, and advertising of products. In addition, it is a marketing tool to steer sales volume of specific products, customize marketing to user, and gain valuable information for market research about the user. It can also contain user-friendly functions for product search and information retrieval [18, p. 512].

EPCs support customers, sales employees, and engineering staff [5, p. 15]. Sometimes EPCs are specialized on one purpose but often a functionality combination is preferred by the implementing companies.

- Customers get information about products, services and conditions offered and configure and order products and services through EPCs.

- These catalogs help sales people to consult the customer, to present and configure products and services to the customer, and to create the appropriate sales documents correctly and on-the-fly. Therefore EPCs are part of a computer aided selling (CAS) approach [13, p. 22].

- Engineering departments of the own company and of business partners are supported by EPCs developing new products or customizing products for piece production.

EPCs, build to serve customers and/or sales people, walk the user through the different initiation phases of ordering a product [25, p. 20]. These phases are:
- Requirements analysis

- Product retrieval

- Product presentation

- Product configuration

Afterwards the customer can order the configured product. Presentation and counsel of subvention and financing possibilities can be included into EPCs [25, pp. 20-21].

At first EPCs have been implemented by companies offering off-the-shelf products. In combination with product configuration tools, electronic product catalogues are also interesting for companies with piece production or series production with many variants. The deployment of electronic product configuration tools helps these companies to decrease the order throughput time. This speed-up and cost-reduction of the production is achieved because product configuration tools check a selected configuration immediately. Thus invalid configurations are rejected – further contacts and inquiries to the customer are eliminated [10].

EPCs are separated into offline and online catalogs. Offline catalogs are usually an electronic version of the common paper mail-order catalogs. These catalogs are easier, cheaper and faster to create than paper catalogs and are shipped to the customer on discs or CDs [14, pp. 8-9]. Product information is stored at these data storage devices. Online catalogs receive product information through communication networks. The presented information is therefore always up-to-date and free of redundancy. Online catalogs become more and more popular because of the spread of the internet and decreasing Internet access costs [2, p. 558]. Obviously, this kind of EPCs is more appropriate in case of a VE for MC. In addition web browsers are interesting clients for product data management systems because of the ability to display all kind of content like text, pictures, tables and other multimedia objects like virtual reality modeling language (VRML) [5, pp. 84-86].

4.2   Production Schedules

Information about the schedule of production orders is of special value in business-to-business (b2b) as well as business-to-consumer (b2c) area. The current status and progress of the production, historic and future (forecast) data are of use for Internet applications.

Historic production schedules are often used during the calculation of project estimations. Especially manufacturers of piece products use historic data to quickly deliver rough time appraisals to sales people and project management. However, historic data is only the second best solution. It contains always an error, cause business environment changes quickly, and information about production capacity, material availability and possible engineering difficulties are not included into the calculation. A simulated scheduling of a production order delivers estimations that are much more reliable. Nonetheless, the use of data from historic production schedules is a quick and cheap way of estimating the needed production time.

Online information about status and progress of production orders is important to create faith in the deliver-reliability of the company especially for the B2C-segment. It is value-added service to inform customers about the state of their order. This is of particular importance for e-commerce sites using a build-to-order approach. Often online customers do not get any type of receipt for their order. A way to check the progress of their orders can help to build a more solid customer-relationship. Safety and security of online shopping are the users' top priorities.

Forecasts about the estimated finish of a production order have a huge importance for a number of applications. In the b2b-relationship this information is needed to manage the supply chain. Depending on the production schedule, suppliers have to deliver the needed materials and the next supply chain partners can schedule their activities.

Many companies use external logistic firms to transport their goods and products to the appropriate customer. Notifying these firms beforehand give them time to reserve the necessary logistical capacities and avoid transportation bottlenecks.

Reliable production data is also required to enable production using a just-in-time (JIT) approach. JIT means to manufacture and supply goods that are needed, when they are needed, and in the exact quantities they are needed [11, p. IV]. Therefore the JIT concept is used to consequently realize the pull approach [8, p. 85]. "The ideal goal of a JIT system is to have the entire production-of-business cycle operate without interruption and without non-value-added time costs." [7, p. 407]. Especially SME need to exchange production data to ensure that the necessary quantity material is made available at the right time without maintaining a huge inventory.

However, production schedule data is also used during the order phase. The MPC system can calculate an estimation finish time. This helps the customer to check, if the stated delivery date meets the timetable of the customer. It builds faith in the producer's ability to keep delivery terms. Reliable estimates for manufactured goods can only be stated if MPC information is included.

One step further lead to the available-to-promise concept. Using logistic and production information ordering customers get specific delivery date promised. Of course, a company can calculate the proposed delivery date with lots of floating time to avoid being wrong. Nevertheless, customers interested in promised delivery dates have usually an important and urgent product order. Therefore, the stated delivery date is indeed a high-priority decision factor for these customers.

Another possibility of Internet application is proposed by Scheer. During long-term production planning companies can use the Internet to cost efficiently survey customers for their expected demand for a specific product. Therefore the customer is asked to fill out an electronic form. This method does not only save distribution costs, but makes also manually data entries obsolete by storing entered information directly into corresponding databases [27, p. 35].

4.3 Capacity Information

The further extension of e-commerce portals and electronic procurement agents will lead to a change of the buying behavior in the b2b segment. These procurement portals offer the capability to automatically procure standardized material at the lowest price available without user participation. A slide change in the pricing or delivery conditions can lead to a huge change of the order volume. A company lowering the price for a specific product could receive so many orders for product within a short period of time, that manufacturing capacity of the company is over-loaded.

Therefore, companies have to include capacity information into their pricing management. A good pricing mechanism can work similar to the yield management used at airlines. Long-dated orders for a time with lots of available capacity are produced at a low price per unit. The unit price increases with less available production capacity during the expected manufacturing time. Surplus manufacturing capacities can be offered at a much lower price, if not booked a few days beforehand. Prices for these last-minute offers can be lowered maximally down to the amount of coverage. This pricing mechanism can be used for the flexible calculation of transfer prices between the participants of the VE.

Another opportunity for the use of capacity information occur at companies with outsourced plant management, which is the typical case in a VE for MC. Maintenance intervals for machinery are usually determined using the estimates of the unit producer. These estimates are founded on the experience of the manufacturer with the appropriate machine under the condition of full working load. Using the capacity information of a MPC system maintenance interval can be extended reflecting the actual wearout of the machinery in question. In addition meter readings and machinery wearout can be gathered and transmitted through the Internet [27, p. 36].

4.4   Other MPC Information

Other information of MPC systems can be as well useful in internet applications. For example, these data includes information of the KANBAN system and workplace schedules.

The KANBAN system is a possibility to implement the JIT concept. In the KANBAN system the procurement of a part is triggered when the KANBAN card related to the demand for this part is set to "empty". The status of KANBAN cards can be seen by supplier through the internet, who would initiate the refill of the production KANBAN [27, p. 36].

Workplace schedule information are especially interesting for company external or internal plant maintenance crews. Plant maintenance is able to determine the optimal point for the inspection of a specific workplace by reviewing the schedule.

5   Approaches to Support Federation of MPC Systems

Specialized software is needed to allow for federation of MPC systems of different companies. Mainly, this software has to provide the basic means for cooperative problem solving. It has to support communication between cooperating MPC systems, and it has to support the cooperation task itself.

Software which supports the mentioned tasks are, e.g. object request broker, transaction manager, workflow management systems, or agent-based approaches [6].

However, to make these approaches work standards are mandatory. Basic standards mostly aim on supporting *electronic data interchange* (EDI). The most important standard for cross-organizational data interchange was established by the United Nations with UN/EDIFACT (electronic data interchange for administration, commerce, and transport) [33]. It standardizes electronic exchange of structured information, e.g. orders, or invoices, thus permitting a direct communication between different business application systems. However, fundamental, economic, and organizational drawbacks causes that UN/EDIFACT did not win the expected recognition and implementation extent.

Open-EDI/object oriented-EDI [31], Universal Data Element Framework (UDEF) [9, pp. 25-26], Basic Semantic Repository (BSR), and its successor BEACON [28] are efforts, that address these problems. They focus on establishing uniform business scenarios and semantic rules. Furthermore, they aim on establishing *business ontologies*. Backward compatibility is supported by repositories, which allow to store relations between known standards, especially UN/EDIFACT, and the newly created ones, cp. e.g. [9, pp. 10-12]. The XML/EDI-Initiative [20] and related efforts [30] on the other hand concentrate on economical and organizational drawbacks of UN/EDIFACT using the extensible markup language (XML) [4] to lower implementation costs and increase flexibility of EDI.

Business data interchange formats and business ontologies are preliminary for federating MPC systems. Based on related standardization efforts, vendors of ERP systems start to specify interfaces to interact with their products, and vendor groups like the Open Applications Group (OAG) try to define interface standards. For example, the OAG defines so-called *Business Object Documents* (BOD). BODs contain syntax and semantics of the date to be exchanged between coarse-grained business functions [20], e.g. manufacturing, financials, or human recourses.

6   Conclusions

In this paper a theoretical approach was presented, how ERP data especially from the field of manufacturing planning and control can be used for the co-ordination of participants of a VE for MC. The data needed were derived from a VE for MC model. Furthermore, technical approaches for the Federation of MPC systems were discussed briefly.

However, the presented ideas were evaluated in a concrete case study with SAP R/3 [30]. All ERP data mentioned above have been extracted from the SAP application server using the SAP Web-Server and data access through BAPI (Business Application Protocol Interface). A first prototype of a simplified Internet ERP system is available now for further research in the field of MC.

References

[1] Arnold, O., Faisst, W., Härtling, M., Sieber, P. (1995): Virtuelle Unternehmen als Unternehmenstyp der Zukunft? HMD 32(185), pp. 8-23.

[2] Birkhofer, H., Büttner, K. (1995): On-line-Produktkataloge. ZWF 90(11), pp. 558-561.

[3] Bray, T., Paoli, J., Sperberg-McQueen, C. M. (Ed.) (1997): Extensible Markup Language (XML). http://www.w3.org/TR/PR-xml.html. Access: 1998-06-12.

[4] Buchner, K., Ullsperger, B. (1995): Elektonische Kataloge im Investitionsgütervertrieb. CIM Management 11(11).

[5] Chu, X., Fan, Y. (1999): Product Data Management based on Web technology. Integrated Manufacturing Systems 10(2), pp. 84-88.

[6] Fellner, K., Turowski, K. (2000): Framework Architecture Enabling an Agent-Based Inter-Company Integration With XML. Australian Journal of Information Systems 8(1), pp. 41-50.

[7] Gunasekaran, A., Lyu, J. (1997): Implementation of just-in-time in a small company: a case study. Production Planning & Control 8(4), pp. 406-412.

[8] Hartland-Swann, J. (1987): MRP or JIT: which is best? In: C. Voss (Ed.): Just in Time Manufacturing: Proceedings of the 2nd International Conference 20-22 October 1987. London, pp. 81-91.

[9] Harvey, B., Hill, D., Schuldt, R., Bryan, M., Thayer, W., Raman, D., Webber, D. (1998): Position Statement on Global Repositories for XML. ftp://www.eccnet.com/pub/xmledi/repos710.zip. Access: 1998-12-01.

[10] Hasse, V. (1998): Produktkonfiguratoren senken Durchlaufzeiten. Logistik heute (12), pp. 75-77.

[11] Hirano, H. (1988): JIT Factory Revolution. Portland.

[12] Houlihan, J. B. (1992): International Supply Chain Management. In: Logistics - The Strategic Issues. Ed. M. Christopher. London, pp. 140-159.

[13] Hüllenkremer, M. (1998): Was leisten Produktkonfiguratoren im CAS-Bereich? Distribution 29(6), pp. 22-23.

[14] Hümmer, M. (1995): Einsatzmöglichkeiten elektronischer Kataloge im Vertrieb. CIM Management 11(11), pp. 6-9.

[15] Kurbel, K., Teubner, A. (1996): Integrating Information-system Development into Business Process Reenginering. In: C. Nagib; C. Callaos (Eds.): Proceedings of the International Conference on Information Systems Analysis and Synthesis. Orlando, pp. 240-246.

[16] Lincke, D.-M., Stanoevska-Slabeva, K., Schmid, B. (1997): Mediating Electronic Product Catalogs. Proceedings of the 7th Mini Euroconference. Bruges.

[17] Mertens, P. (1995): Mass Customization (Massen-Maßfertigung). Wirtschaftsinformatik 37(5), pp. 503-506.

[18] Mertens, P., Lödel, D., Thesmann, S., Breuker, J.-S., Ponader, M., Kohl, A., Büttel-Dietsch (1992): Elektronische Produktkataloge - Entwicklungsstand und Einsatzmöglichkeiten. Wirtschaftsinformatik 34(5), pp. 509-516.

[19] Nagel, R., Dove, R., Preiss, K. (1991): 21st Century Manufacturing Enterprise Strategy - an Industry led View. Vol. 1, Bethlehem, PA.

[20] OAG (Ed.) (1997): White Paper: Open Applications Integration: Projects of the Open Applications Group.

[21] Peat, B., Webber, D. (1997): Introducing XML/EDI: "The E-business Framework". http://www.geocities.com/WallStreet/Floor/5815/start.htm. Access: 1998-12-01.

[22] Piller, F. (1998): Kundenindividuelle Massenproduktion: Die Wettbewerbsstrategie der Zukunft. München.

[23] Pine II, B. J., Victor, B., Boynton, A. C. (1993): Making Mass Customization Work. Havard Business Review 36(5), pp. 108-119.

[24] Pine II, J. B. (1993): Mass Customization: The New Frontier in Business Competition. Boston.

[25] Popp, H., Schumann, P. (1995): Electronic Shopping im Business-to-Business Bereich. CIM Management 11(11), pp. 18-22.

[26] Rautenstrauch, C. (1997): An Alternative Concept to MRP II for Mass Customization Based on the Object-oriented Paradigm. In: S. T. Tan; T. N. Wong; I. Gibson (Eds.): International Conference on Manufacturing Automation. Vol. 1, Hong Kong, pp. 401-406.

[27] Scheer, A.-W., Bold, M., Hoffmann, M. (1998): Internet-basierte Geschäftsprozesse mit Standardsoftware. In: SAP R/3 in der Praxis. Eds.: D. Preßmar; A.-W. Scheer. Wiesbaden,.

[28] Steel, K. (1997): The Beacon User's Guide: Open Standards for Business Systems. http://www.cs.mu.oz.au/research/icaris/beaug1.doc. Access: 1998-12-01.

[29] Steffen, T. (2000): Internet-Quellen zu XML/EDI. Wirtschaftsinformatik 42(1), pp. 78-86.

[30] Tangermann, H. (2000): Internet-based Manufacturing Planning and Control Systems using SAP R/3. Diplomarbeit. Magdeburg.

[31] TMWG (Ed.) (1998): Reference Guide: "The Next Generation of UN/EDIFACT": An Open- EDI Approach Using UML Models & OOT (Revision 12). http://www.harbinger.com/resource/klaus/tmwg/TM010R1.PDF. Access: 1998-12-01.

[32] Turowski, K. (2000): Supporting Innovative Competitive Strategies As Mass Customization By Pairing E-Commerce Techniques With Agent Technology. In: Electronic Commerce: Opportunities and Challenges. Eds.: S. M. Rahman; M. Raisinghani. Hershey, pp. 261-278.

[33] UN (Ed.) (1995): United Nations Directories for Electronic Data Interchange for Administration, Commerce and Transport. http://www.unece.org/trade/untdid/Welcome.html. Access: 1998-12-01

Customer Interaction and Digitizability – a Structural Approach to Mass Customization

Piller, F. T.

Summary: Enterprises in all branches of industry are being forced to react to the growing individualization of demand, yet, at the same time, increasing competitive pressure dictates that costs must also continue to decrease. Companies have to adopt strategies which embrace both a closer reaction to the customers' needs and efficiency. Mass Customization meets this challenge by offering individually customized goods and services at mass production efficiency. However, while Mass Customization has already been discussed in the literature for more than a decade, increased practical implementation of this strategy can been found in business only within the last years. This time lag may be explained by the fact that only since few years sufficient technologies exist to handle the information flows connected with mass customization. Especially as mass customization enters more and more consumer markets, new Internet technologies can be seen as its main enabler. To connect strategies discussed in e-business with the field of mass customization, the paper deploys a structural approach to create mass customized concepts within electronic business.

1 Individualization, Competitive Pressure, and Mass Customization

„It is the customer who determines what a business is." In the very sense of Drucker's (1954) analysis, the single customer has come more deeply into the firm's focus than ever. Firms are faced by an uninterrupted trend towards individualization in all areas of life, as new Delphi studies predict. Explanations may be found in the tendency towards an experience economy, the growing number of single households, an orientation towards design and, most importantly, a new awareness of quality and functionality which demands durable and reliable products corresponding exactly to the specific needs of the purchaser. In particular, consumers with great purchasing power are increasingly attempting to express their personality by means of an individual product choice (an example is BMW's new „Individual Program" which emphasizes the fulfillment of individual fittings and equipment). Thus, many suppliers are forced to create product programs with an increasing wealth of variants right down to the production of units of one (differentiation by means of variety). In the final consequence, many companies have to process their customers individually (Glazer, 1999; Kahn, 1998).

Traditionally, the objective of customizing goods and services is to attain an increased revenue by the ability to charge premium prices derived from the added value of a solution meeting the specific needs of a customer (Porter, 1980). However, the present competitive situation of many industries prevents the company from reacting by a strategy of differentiation. The cost-benefit relation alters because buyers demand relatively high standards of quality, service, variety or functionality even when the sales price is favorable or, vice versa, suppliers have to meet additional requirements in pricing when a product is marketed differentiated.

Companies operating successfully under this condition have to start out from both the cost as well as the differentiation position at the same time. However, this is in conflict with the classical generic competitive strategies according to Porter (1980) – i.e. *differentiation, cost leadership, and focus strategies* – which became the precepts on which countless companies based their operations (Kotha, 1995). In his conception, Porter follows the fundamental postulate of the incompatibility of differentiation *and* cost leadership. A company must clearly decide on one type of strategy, otherwise it runs the risk of getting "stuck in the middle" (Porter, 1980:16). However, empirical studies (e.g. Kekre/ Srinivasan, 1990; Miller/Dess, 1993; Reitsperger et al., 1993) and a detailed theoretical argumentation (e.g. Faulkner/Bowman, 1992; Hill, 1988; Murray, 1988) demonstrate that competitive strategy does not necessitate choosing between cost leadership *or* differentiation. Rather the simultaneous attainment of both strategic positions should be pursued within the context of a hybrid competitive strategy.

The practical implementation of hybrid competitive strategies is based on the potential offered by new technologies in manufacturing and information management. At the time Porter's conception came about, process technologies that are now perfected were only in the stage of development. New manufacturing technologies (computer-integrated production and flexible manufacturing systems) reduce the trade-off between a wide range of variants (flexibility) and production cost (productivity). In many cases, however, the essential prerequisite for the implementation of a hybrid strategy is the electronic networking of purchasers and producers as well as the suppliers involved. Adequate technologies are available nowadays with the Internet and its sub-technologies (like SCM, XML, VPN and so on).

Precisely this combination of strategic challenges and new technological possibilities is the driver of mass customization. While Davis coined the term in 1987, the concept attained wide popularity with Pine's (1993) book. Mass customization means the production of goods and services for a (relatively) large market that exactly meets the needs of every individual demander with regard to certain product characteristics (*differentiation option*) at costs roughly corresponding to those of standard mass-produced goods (*cost option*). The information collected in the course of the process of individualization serves to build up a lasting individual relationship with each customer (*relationship option*). In order to focus the discussion, the remaining paper concentrates on the mass customization of goods (and not services). Further, the term 'customer' always refers to the end customer, i.e. the consumer or user of the customized product.

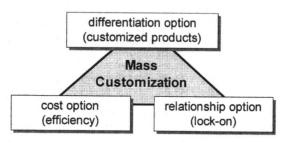

Figure 1: The three levels of mass customization

The implementation of mass customization takes place by means of various methods, which combine different options for customization while maintaining the cost option (similar classifications are described by Gilmore/Pine, 1997; Lampel/Mintzberg, 1996; McCutcheon et al., 1994; Piller/Schoder, 1999; Pine, 1993; Robertson/Ulrich, 1998). Figure 2 shows that modularization (*platform thinking*) can be regarded as the central principle of mass customization (Feitzinger/Lee, 1997; Pine, 1993:196-212; Sawhney, 1998): "A product with a modular design provides a supply network with the flexibility that it requires to customize a product quickly and inexpensively" (Feitzinger/Lee, 1997: 117). A relatively high level of prefabrication permits scale-oriented basic production whose results (modules) are combined or completed in the final steps of production according to a specific customer order.

Soft Customization: Customization based on fully standardized manufacturing processes	Hard Customization: Customization starts within the manufacturing processes	Degree of customer specific activities of the value chain
Self customization create customizable products and services *Microsoft Office, Aveda Personal Blends, Lutron Electronics*	**Customization-Standardization-Mix** either the first or the last activities of the value chain are customized within the factory, while keeping the others standardized *Personal Spin (Levi Strauss), IC3D, MySki*	
Point-of-delivery customization customization of a standardized product at the point-of-delivery *Paris Miki, Dynafit and Nordica ski shoes*	**Modular product architectures** modularize components and combine them to customized products *Anderson Windows, Ross Controls, Dell, Creo-shoes, Idtown, Customatix*	
Service customization customize services around standardized products and services *1-800-Birthday, Hertz Gold Club, Peapod*	**Flexible customization** using flexible manufacturing systems for production of fully customized products without higher costs *Sandvik Coromant, Sovital, Reflect*	

for more examples of mass customization see www.mass-customization.de

Figure 2: Methods for achieving mass customization

From a strategic point of view, mass customization means differentiation through customization, i.e. the production of goods in so many variants that the wishes of each relevant customer are fulfilled. While most variety strategies assume that goods are produced in advance for defined market niches and placed in inventory for some anonymous customer, a mass customized product is manufactured individually for an identified customer after the order has been received. The demand for a relatively favorable cost level constitutes the cost option of mass customization. A customized product can lead to lower costs when a combination of economies of scale and economies of scope is attained („economies of integration", Noori, 1990). The differentiation option leads to greater product attractiveness. In addition, the individual contact between supplier and customer offers possibilities of building up a lasting relationship with the customer ("learning relationships", Peppers/Rogers, 1997). Once the customer has successfully purchased an individual item, the knowledge acquired by the supplier during the product configuration represents a considerable barrier against switching suppliers. Even if a competitor possesses the same mass customization skills and even if he offers a lower price, a switching customer would have to go again through the procedure of supplying information for product customization. Also, she is once again faced with uncertainties in regard to the quality and the producer's behavior. As these two factors lead directly to an increase in sales, production volumes are increasing, too – the basis of economies of scale –, while modular product architectures allow firms to attain economies of scope.

At the same time, the individual production of goods results in *economies of efficiency* (better planning conditions, reduction of fashion risks, reduction of stock keeping of goods for distribution, see Piller 2001 for a detailed discussion). When the information acquired by the company about its various customers is aggregated and compared, customer behavior becomes transparent (Kotha, 1995). New customers can be served better and more efficiently, because they are offered an individual product variation which other customers with a similar profile have already purchased in the past ("profiling"). In addition, the broad information basis allows firms to cut back on pools of fixed costs that came about due to the necessity of maintaining a high level of operational flexibility.

2 Mass Customization and Electronic Business

Until today, mass customization is connected closely to the potential offered by new manufacturing technologies (CIM, flexible manufacturing systems) reducing the trade-off between variety and productivity (Ahlström/Westbrook, 1999; Anderson, 1997; Kotha, 1995; Pine, 1993; Rautenstrauch, 1997; Tseng, 1997; Victor/Boynton, 1998; Zipkin, 2001). However, we want to argue that information shall be regarded as the most important factor for the implementation of mass customization. „Being truly customer focused is not possible if the organization is not, first, information intensive" (Blattberg/Glazer, 1994). As shown in Figure 2, there are different conceptions to implement mass customization with diverse demands on production, all methods lead towards a sharp increase in the amount of information and communication necessitated among those involved. Mass customization is successful only when it can cover this need for information and communication both purposefully *and* efficiently.

The reason for this information richness is based – in comparison to the traditional push-system of mass production – on the need for direct interaction between the customer and seller for every single transaction, a mechanism that will be discussed more detailed at the end of this paper. Every order implies coordination about the customer specific product design as a result of the divided construction process of mass customization (Hibbard, 1999). While the product architectures and the range of possible variety are fixed during a preliminary design process the second step takes place in close interaction between the customer and the supplier. The individual wishes and needs of each customer have to be transformed to a unique product specification. The costs arising from customization consist largely of information costs. They are accounted for by the investigation and specification of the customers' wishes, the configuration of individual products, the transfer of the specifications to manufacturing, an increased complexity in production planning and control, the coordination with the suppliers involved in the individual prefabrication, and the direct distribution of the goods. All theses activities are characterized by a high information intense compared to traditional mass production. Thus, customer-related value added is produced on the information level.

The importance of information processing for mass customization may explain the observation that most prominent examples of mass customization were founded just within the last couple of years – although the concept has already been discussed in management literature for more than a decade (e.g. Davis, 1987; Kotler, 1989; Pine; 1993; already Toffler, 1970 described the basic idea). Explanation for that time lag may be found in the fact that only within the last years sufficient technologies exist to handle the information flows connected with mass customization. In former times, firms reduced the information content of their processes in order to reach cost efficient outputs. But today the opposite can be true: An increasing information richness of products and processes guarantees its cost efficient and individualized production. This is possible through the potentials of new information technologies (Wigand/Picot/Reichwald, 1997).

Especially as mass customization enters more and more consumer markets, new Internet technologies can be seen as its main enabler. While in business-to-business markets personal sale and configuration is common, in consumer markets the interaction often has to be fulfilled over the Internet. Web-based interaction tools like product configurators allow to „outsource" the time- and cost-consuming configuration process to the customer. For low cost consumer goods extensive sales and configuration processes cannot be fulfilled by personal sales in a retail outlet – if keeping the cost option of mass customization in mind. Customized cosmetics like reflect.com with a retail value of 10 € cannot be sold in traditional channels. The same is true for many other mass customized products with a relatively small margin. Therefore, mass customization can be seen closely related to e-business and new possibilities connected with the Internet economy. The use of the Internet as a communication medium facilitates the efficient production of customized goods as well as the personalization of customer relationships. Therefore, the vital role of information categorizes mass customization from a conceptual point as an application of Electronic Commerce respectively an e-Business strategy (Duray et al., 2000; Fulkerson/Shank, 2000; Lee/Barua/Whinston, 2000; Reichwald/Piller/Möslein, 2000b; Zerdick et al., 2000).

3 The Information Cycle of Mass Customization

Our research showed that companies pursuing mass customization successfully integrate a variety of important tasks (for the empirical word building the background for this paper see Reichwald/Piller/Moeslein 2000a, 2000b). They build an integrated information flow – that not only covers one transaction but also uses information gathered during the fulfillment of a customer-specific order to improve their knowledge base. The representation of these processes and of the tasks described before in an information cycle model shall stress the importance of an interconnected and integrated flow of information (Figure 3).

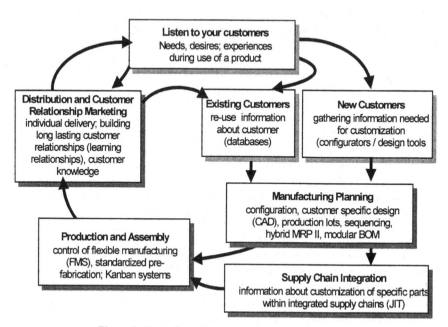

Figure 3: The information cycle of mass customization

1. Listen to your customers: The cycle starts with the individual needs of each customer. The center of each mass customization program has to be information about the desires of a customer group regarding the product. Although a major meaning of mass customization is that a mass customizer doesn't fulfill every wish of its customers (that would be traditional customization at premier prices), it is important to listen carefully to prospective customers to design a set of product variants and individualization options that on the one hand side has enough possibilities for customization, but that on the other hand is as easy as possible in order to reduce complexity – a main cost driver of mass customization.

2. Configuration: Here the task is to transfer the customers' wishes in concrete product specifications. This is one of the most critical parts of any mass customization business. It's important to differentiate between old and new customers. For new customers, first a general profile of their desires and wishes has to be built up. This profile is transformed into a product specification. At this stage new technologies like recommendation engines provide help. For the configuration for regular customers the existing customer profile has to be used. The old configuration may be presented and customers just asked for variations. The second and all following sales have to be as easy (time- and money-saving) as possible. This is one of the major possibilities to built-up customer loyalty. Leading companies have implemented strong instruments to build trust and show reliability in order to reduce the risk seen by prospective customers in mass customization processes („configure your own design, pay first, wait, and then hope, that our product fit", a sales manager described the transaction process from a customer's point of view). New research by Mandel/Johnson (1999) demonstrates strong possibilities to influence the users of a Web-site by screen design. These findings have to be used to develop „trust-full" Web-sites.

3. Manufacturing planning: Often already during step 2 the configuration is checked in production planning to get customer specific dates for delivery. After an order is placed, it is transferred into specific manufacturing tasks. Scheduling activities follow. The production tasks are transferred to the responsible process units, whereby suppliers may be integrated in the customization of some parts, too. Note that there shall be no step back to the configuration process after the order was placed. This time- and money-consuming iterative re-configuration has to be eliminated in a successful mass customization concept.

4. Production and supply chain integration: Up to this point, the mass customization process is on the information level only. Now manufacturing activities starts. Often in a segmented production layout (one production segment is responsible for some modular product components) the order is fulfilled. During this step information management has to take care that the right specification of an order are at the right work place at the right time. In an advanced mass customization concept, external suppliers may be integrated into the customization process. This allows to extend the economically possible degree of individualization, a speeding up of the processes, and cost savings due to specialization and faster learning effects. Here, too, information activities are in the center. Integrated information flows and shared applications are required to transfer the customer specific information between the factories.

5. *Relationship management:* After distribution, the relationship building continued that started with the configuration process. Further knowledge about the customers has to be acquired. The information cycle also shows that not only information about the customer but the production process itself has to be collected in a knowledge base to improve efficiency and quality in follow-up business (Peppers/Rogers, 1997). By doing so, not only new and old customers can be served better. Also production planning can be improved continuously (e.g. by better forecasts for the prefabrication of modular components).

4 Systematization of Mass Customization Concepts

While all companies pursuing mass customization successfully have to build an integrated information flow, there is not a single first-best solution for mass customization. We showed already above that there are different approaches to address mass customization. However, this differentiation of mass customization concepts arguments from a totally manufacturing orientated perspective (like all other know structural approaches of mass customization, for an overview see Piller 2001). In the remaining paper, we will present an approach to structure mass customization strategies that addresses two distinctive features of the concept: the interaction with the customer and the importance of information and electronic business. Thus, for four strategic fields can be identified which are presented in Figure 4. The strategic tasks and demands of each of these fields will be discussed in the next chapter. Before we want to have a closer look on the structural variables, the *degree of interaction required* between the customer and the manufacturer and the *degree of digitizability* of customized product components.

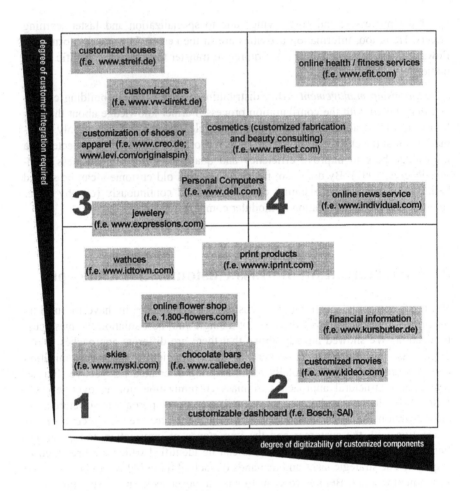

Figure 4: Mass customization – a structural approach

Degree of interaction required: The integration of the customer into the production process is a distinctive feature of customized production. One can look at the relation between the customer and supplier as a cooperation providing benefits for both sides, but demanding inputs of both participants, too. In mass customization processes, this integration of the customer is required primarily during configuration. The "costs" of this process from the customer's point of view are an important success factor. Especially in consumer markets customers often do not have sufficient knowledge for the definition of the product specification, which corresponds to their needs. As a result the configuration process may last pretty long, and customers may experience an increasing uncertainty during the transaction process. Comparison processes are more difficult because of smaller transparency of supply compared to standardized goods or services. Uncertainty exists, too, about the behavior of the provider. Further, the cooperative character of the configuration results in an asymmetrical distribution of information – a typical principal agent constellation. Information gaps are increasing the newer and more complex individualization possibilities are. Without a clear reference point for the definition of an optimal performance it is also difficult to judge whether a case of warranty arose.

These uncertainties and factor transfers can be interpreted as additional transaction costs of the customer. One of the most important tasks of the mass customizer it to ensure that the customer's expenditure is kept as small as possible, while the benefit she experiences has to be clearly perceptible. Leading companies of our sample have implemented strong instruments to build trust and reliability in order to reduce the risk seen by prospective customers in mass customization processes. Other instruments minimizing the risk of the customer are warranties or the reputation of the provider. But independent from trust and warranties, the degree of customer integration required into the customization process is positively connected with the expenditures and risk realized by the customer. Thus, the degree of interaction required for customization will not be equal for all products and services. The buyer of a personalized gift watch of *idtown.com* with a purchase price of 35 € will experience smaller complexity of the purchase process than the buyer of a *VW* sedan, which can be configured and ordered completely without engagement of a dealer on the Web Site of *Volkswagen-Direct*. Therefore, the characteristics of the product or service being individualized have to be taken into account. Accordingly, the degree of customer integration required is influenced by the relative price of the products and services, the possibility to use instruments to prevent bad investments (e.g. warranties, exchange policy, time of delivery, screening possibilities), the customer's experience with a product (e.g. second buy, product specific knowledge) and its complexity (customization possibilities; product structure). Another point to consider is if the configuration process itself can be regarded as part of the product purchased (configuration as buying experience, leisure activity).

Degree of digitizability: While all mass customized products are characterized by a more or less intense integration of the customer, not all of them can be transferred to an „info sphere" where the customization of products and services can be delivered often very efficiently. The traditional mass producer reduced the information intensity of his products and production processes as far as possible to achieve a good cost position. However, today a fundamental enabler for an economical individualization of products and services is to increase their information intensity. New information technologies allow to substitute product functions by information activities. Information richness is a strong indicator for the digitizability of goods and services (Shapiro/Varian, 1998). *Digitizability* describes the extent in which functions relevant to a customer can be fulfilled by the use of information technology only. For fully digital products modern information technology is the only necessary instrument to deliver customization. These goods can be sold, customized, and distributed over computer networks relatively easily and without high additional costs – often regarded as the "heart" of Electronic Business (Choi/Stahl/Whinston 1997). The degree of digitizability is based on the content of information based components in relation to the whole product or service. While products like newspapers or music can be digitalized totally others like apparel have an information content of zero. But also for this products new Internet technologies facilitate the collection and employment of numerous data concerning the individual customer by permitting interaction between economic units connected via electronic networks.

5 Four Strategies to Create Competitive Advantage With Customized Goods and Services

Based on different combinations of possible degrees of interaction required and digitizability four fields can be derived, representing different business models of mass customization.

(1) Add-on: A first group of rather simple, not complex products and services shows a very small or no degree of digitizability. Also, there is no deep interaction necessary to collect the information required for customization as only few components can be individualized or most customers have enough knowledge about the product so that they can find the sufficient configuration easily and feel no large risk in the buying process. The production of these products is based on the potentials of modern manufacturing and logistics systems. However, the configuration (interaction with the customers) and the planning of the operations are based on integrated information flows. Success factors for theses products and services are first integrated information flows connecting the production system with configuration. Second, the degree of interaction has to be increased in order create a deeper learning relationship with the customer (as long as this can be automated from the supplier's point of view). Therefore, new functionalities or additional online services may be added. By doing so, the degree of digitizability can be also increased by providing easy-to-implement customized services around a (standardized) core product. Although the core product of an online flower shop (just in time delivery of a bouquet) is not digitizable, it differentiates itself from brick-and-mortar stores – and other competitors on the web – in particular through customized services such as a birthday reminder, an address book or a writing program for creative greeting cards.

(2) Attract attention: While the degree of customer integration required is small in this group, customization can be digitized to a large extent. These products and services are information products in the broadest sense that can be sold, custom-

ized and delivered via e-business-networks. Customization serves primarily to differentiate a company form its competitors in order to increase customer loyalty. Mass customization can be seen as a strategy to create economics of attention (Goldhaber, 1997; Shapiro/Varian, 1998). Economics of attention are a result from the observation that information is freely available on the Internet while the attention of the users is limited. By increasing the degree of attention, a company shall be able to introduce new products and thus to implement network effects more easily. In this context, mass customization offers an additional way to attract attention based on the incentive of additional customer benefits. While, for example, today some thousand Internet radio stations compete for the attention of the users, *imagineradio.com* allows its listeners to create their very own radio station playing their preferred music. Here, mass customization serves as a base for new models to create economic value: As the high degree of digitizability makes customization economically feasible, the individualized product or service is often offered free of charge. The objective is to initiate a learning relationship and to gain more information about the individual customers in order to sell either (standardized) products and services fitting to the customized information content or just advertising space on the web site.

(3) Configuration: The third group is characterized by the requirement of an intensive interaction with the customers. An example may be customized apparel where personal measurement activities are necessary during the configuration process. In mass markets, this has to been done by three-dimensional-scanning devices if a company wants to get reliable measurements economically. But body scanning can not be done online – a personal interaction is needed. The same is true for relatively expensive or complex (from the customer's point of view) products and services, for example the customization of diamond jewelry or buildings. A customer may not have the necessary know-how to define a configuration corresponding to her desires. An intensive interaction is also needed from the seller's point of view in order to create confidence and to minimize the purchase risk. Often a multi-channel sales strategy is recommendable to fulfill the different preferences of diverse customer groups in regard to online literacy, time sensitivity or product knowledge. Several direct and indirect sales forms can be combined – from direct sales by call centers over self-service on the Internet up to brick-and mortar stores. The task of traditional dealers switches hereby from the selling of a product or service to the configuration support and the consultation of the customer. As often a repurchase based on an available customer profile is handled more economically online, new profit schemes for the dealers have to be established (e.g. a higher initial bonus and smaller additional provisions for every online sale). After configuration, customer data have to be transferred into the producer's business application systems.

While the customization itself is not digitizable, electronic business offers an extensive support of the transaction phase. Companies should try to fulfill as much of the configuration process online, but now (compared to group 1) human advise should be offered by a help or call back button. New Internet technologies offer plenty possibilities – from 3D-pictures of the configured objects up to chat robots allowing a conservation in natural speech. Software tools like recommendation engines simplify the identification of preferences by recording, comparing, and aggregating former sales, pages views or click rates. They enable the direct presentation of individualized content and offer a first suggestion of a configuration by comparing user profiles and indexes of content – even if a user cannot explicitly express her preferences and wishes (Elofson/Robinson, 1998; Shardanand/Maes, 1995). These and other technologies enlarge the range of „configure-it-yourself" to more complex products. Thus, even complex products like houses can be configured and ordered online today. The saving potential of this process allows German building contractor *Streif AG* to offer customers a discount of 25 000 € if they configure and order their house online (www.streif.de). The use of these technologies has to be guided by the strong necessity to show reliability and to built trust to reduce the purchasing risk from the customer's point of view. New research by Mandel/Johnson (1999) demonstrates strong possibilities to influence the users of a web site by screen design. These findings have to be used to develop „trust-full" web sites. The whole configuration and interaction process – either online or offline – has to differentiate between old and new customers. For new customers, a general profile of their desires and wishes has to be built up using the technologies mentioned above. For existing customers the old configuration and information gathered during former transactions have to be used to make all following sales as easy (time- and money-saving) as possible. For example, the last configuration may be presented and customers just asked for variations. This is one of the major possibilities to built-up customer loyalty.

(4) E-Service-Innovations: The fourth group is characterized by a high degree of interaction and digitizability. Examples are information goods like consulting and information services. An online health center may replace the work of (expensive) wellness consultants and trainers. After an initial health check each day a customized "wellness map" is created and distributed electronically covering a plan for the daily workout, nutrition suggestions and perhaps the calculation of today's bio-rhythm. Users have to report daily several feedback data. Thus these services must offer a high benefit for a customer to make this expenditure worthwhile. The customer interaction serves as an excellent base to establish intensive learning relationships. The mass customizer has to use information about the individual customer to serve her more comfortably (i.e. at smaller interaction expenditure) and to deliver improved personalized content. That's the starting point for new cross-selling activities. The information about the fitness condition of a customer may provide the initial configuration for customized vitamin products or skin treatment (that strong trust and security issues are essential has not be stressed further).

Services of this group offer a substantial potential for price distinction and personalized pricing – one of the major strategies suggested in e-business. This is done normally by versioning, a buyer chooses under different given versions the product version corresponding to her personal ability to pay (Shapiro/Varian 1998). Mass customization allows to switch from versioning (selection between fixed, bundled products) to an individual pricing system based on different valued components that can be mixed freely. Beyond that, individualized digital products reduce the incentive for buyers for arbitrage businesses between users that had to pay different prices. A further source of income is the aggregation of the individual customer information to customer know-how, which can be sold as market research to other companies. The customizable news services for music lovers *mylaunch.com* or the on-line supermarket *Peapod* have a substantial source of income from this aggregated customer knowledge (reaction to price adjustments, advertising measures, reaction on new products).

6 Conclusion

The mass customization landscape today reveals a somewhat sobering picture. The opportunities of mass customization are acknowledged as fundamentally positive by theory and empirical studies for many years. They have attained a lot of attention by managers from all kind off industry within the last years. Many companies are already successfully operating after this new business model. Yet the concept of electronic business based on the technological potential offered by new information and communication technologies represents the ideal foundation for providing customized products without a significant increase in costs even in mass markets and for exploiting opportunities to establish profitable long-term customer relationships efficiently. Mass customization must be included to a greater extent as a serious option in corporate strategic considerations in order to meet the new competitive challenges. Likewise, opportunities are being thrown away by companies that, by interpreting mass customization as no more than a new trend that should not be missed, merely provide some of their customer interfaces with a veneer of innovative and individual measures. In doing so they fail to change the entire value chain in an integrated manner towards the customer. But only if the information cycle of mass customization will be followed along the whole supply chain network – including the customer –, all the advantages of mass customization will come and stay alive.

While often mass customization is discussed in connection with new manufacturing technologies, we argued in this paper that it is information that is the main enabler and therefore that it are new information technologies that are the major driver of mass customization. To address this point, we deployed a new approach to structure different concepts of mass customization – all with there own distinctive demands and requirements. This argumentation is summarized in Figure 5. However, the suggestions for action found there must not be understand as generic strategic patterns but rather as ideas where a successful mass customization concept can start.

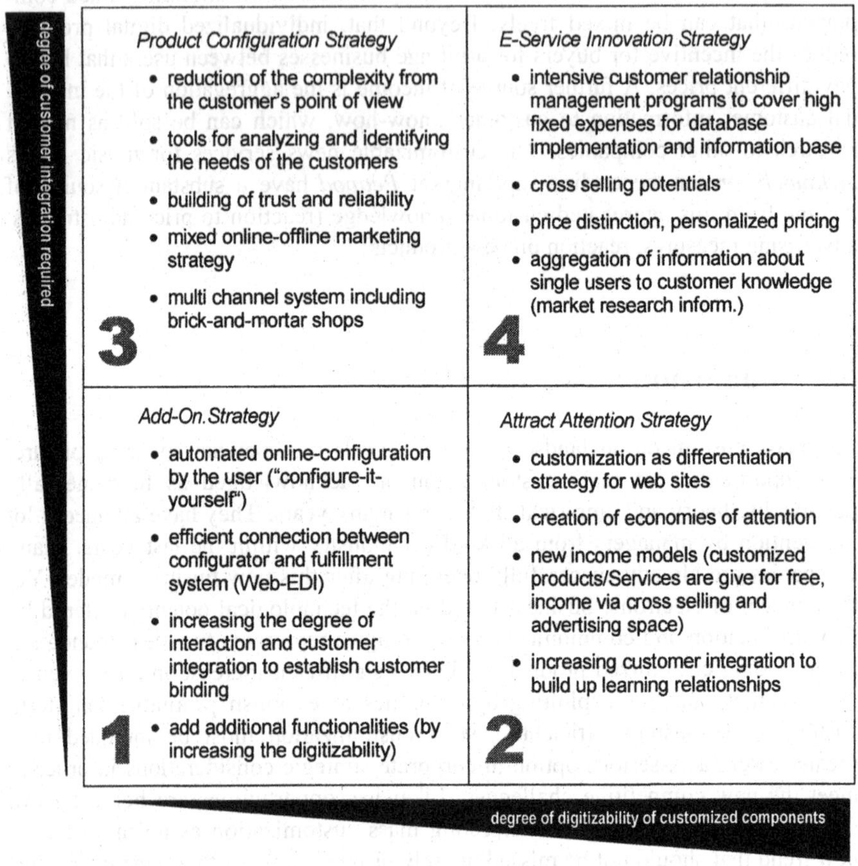

<table>
<tr><td rowspan="2" style="writing-mode: vertical">degree of customer integration required</td><td>

Product Configuration Strategy

- reduction of the complexity from the customer's point of view
- tools for analyzing and identifying the needs of the customers
- building of trust and reliability
- mixed online-offline marketing strategy
- multi channel system including brick-and-mortar shops

3

</td><td>

E-Service Innovation Strategy

- intensive customer relationship management programs to cover high fixed expenses for database implementation and information base
- cross selling potentials
- price distinction, personalized pricing
- aggregation of information about single users to customer knowledge (market research inform.)

4

</td></tr>
<tr><td>

Add-On.Strategy

- automated online-configuration by the user ("configure-it-yourself")
- efficient connection between configurator and fulfillment system (Web-EDI)
- increasing the degree of interaction and customer integration to establish customer binding
- add additional functionalities (by increasing the digitizability)

1

</td><td>

Attract Attention Strategy

- customization as differentiation strategy for web sites
- creation of economies of attention
- new income models (customized products/Services are give for free, income via cross selling and advertising space)
- increasing customer integration to build up learning relationships

2

</td></tr>
</table>

degree of digitizability of customized components

Figure 5: Mass Customization strategies in e-business

References

Ahlström, P. and Westbrook, R. (1999): Implications of mass customization for operations management: an exploratory survey, International Journal of Operations&Production Management, 19 (March 1999), pp. 262-274.

Anderson, D.M. (1997): Agile product development for mass customization, Chicago: Irwin, 1997.

Blattberg, R.C. and Glazer, R. (1994): Marketing in the Information Revolution, The Marketing Information Revolution, R.C. Blattberg et al. (eds.), Boston: Harvard Business School Press, 1994, pp. 9-29.

Choi, S.-Y., Stahl, D.O. and Whinston, A.B. (1997): The Economics of Electronic Commerce, Indianapolis: Macmillan Technical Publ., 1997.

Davis, S. (1987): Future Perfect, Reading: Addison-Wesley, 1987.

Drucker, P.F. (1954): The Practice of Management, New York: Harper, 1954.

Duray, Rebecca et al. (2000): Approaches to mass customization: configurations and empirical validation, in: Journal of Operations Managements, 18. Jg. (2000), S. 605-625.

Elofson, G. and Robinson, W.N. (1998): Creating a custom mass-production channel on the Internet, Communications of the ACM, 41 (March 1998), pp. 56-62.

Faulkner, D. and C. Bowman (1992): Generic strategies and congruent organizational structures, European Management Journal, 10 (No. 4, 1992), pp. 494-499.

Feitzinger, E. and Lee, H. (1997): Mass customization at Hewlett-Packard: the power of postponement, Harvard Business Review, 75 (No. 1, 1997), pp. 116-121.

Fulkerson, B. and Shank, M. (2000): The new economy electronic commerce – and the rise of mass customization, in: Michael Shaw et al. (Hg.): Handbook on electronic commerce, Berlin u.a. 2000, S. 411-430.

Gilmore, J.H. and Pine, B.J. (1997): The four faces of mass customization, Harvard Business Review, 75 (No. 1, 1997), pp. 91-101.

Glazer, R. (1999): Winning in smart markets, Sloan Management Review, 40 (Summer 1999), pp. 59-69.

Goldhaber, M.H. (1997): "Attention Shoppers!," Wired Magazine, 1997, No. 12.

Hibbard, J. (1999): Assembly online, Information Week, 12 April 1999, pp. 85-86.

Hill, C.W. (1988): Differentiation vs. low cost or differentiation and low cost, Academy of Management Review, 13 (No. 3, 1988), pp. 401-412.

Kahn, B.E. (1998): Dynamic relationships with customers: high-variety strategies, Journal of the Academy of Marketing Science, 26 (Winter 1998), pp. 45-53

Kekre, S. and K. Srinivasan (1990): Broader product line, Management Science, 36 (No. 10, 1990), pp. 1216-1231.

Kotha, S. (1995): Mass customization: implementing the emerging paradigm for competitive advantage, Strategic Management Journal, 16 (special issue 'Technological transformation and the new competitive landscape', 1995), pp. 21-42.

Kotler, P. (1989): From mass marketing to mass customization, Planning Review, 18 (No. 5, 1989), pp. 10-13, 47.

Lampel, J. and Mintzberg, H. (1996): Customizing customization, Sloan Management Review, 37 (Spring 1996), pp. 21-30.

Lee, C.-H., Barua, A. and Whinston, A. (2000): The Complementarity of Mass Customization and Electronic Commerce, in: Economics of Innovation and New Technology, 9. Jg. (2000), H. 2, S. 81-110.

Mandel, N. and Johnson, E.J. (1999): Constructing preferences online, Working Paper, The Wharton School of Business, University of Pennsylvania, 1999.

McCutcheon, D. et al. (1994): The customization-responsiveness squeeze, Sloan Management Review, 35 (Winter 1994), pp. 89-99.

Miller, A. and G.G. Dess (1993): Assessing Porter's model in terms of its gereralizability, accuracy and simplicity, Journal of Management Studies, 30 (No. 4, 1993), pp. 553-585.

Murray, A. (1988): A contingency view of Porter's 'generic strategies', Academy of Management Review, 13 (No. 3, 1988), pp. 390-400.

Peppers, D. and Rogers, M. (1997): Enterprise One to One, New York: Doubleday, 1997.

Piller, F. (2001): Mass Customization, 2nd edition, Wiesbaden: Gabler 1999.

Piller, F. and Schoder (1999), D.: Mass Customization and Electronic Commerce, Zeitschrift fuer Betriebswirtschaft, 69 (October 1999), pp. 1111-1136.

Pine, B.J. (1991): Paradigm shift: From mass production to mass customization, Master thesis, Cambridge: Massachusetts Institute of Technology, 1991.

Pine, B.J. (1993): Mass Customization, Boston: Harvard Business School Press, 1993.

Pine, B.J., Victor, B. and Boynton, A. (1993): Making mass customization work, Harvard Business Review, 71 (No. 5, 1993), pp. 108-119.

Porter, M.E. (1980): Competitive Strategy, New York: The Free Press, 1980.

Rautenstrauch, C. (1997): An alternative concept to MRP II for mass customization, Proceedings of the International Conference of Manufacture Automation 1997, The University of Hong Kong, pp. 401-406.

Reichwald, R., Piller, F. and Möslein, K. (2000a): Information as a critical success factor for mass customization, Proceedings of the ASAC-IFSAM 2000 Conference, Montreal, 2000.

Reichwald, R., Piller, F. and Möslein, K. (2000b): Mass Customization Based E-Business Strategies, Proceedings of the Strategic Management Society (SMS) 20th International Conference 2000, Vancouver, British Columbia, Canada 2000.

Reitsperger, W. et al. (1993): Product quality and cost leadership: compatible strategies, Management International Review, 33 (extra issue No.1, 1993), pp. 7-21.

Robertson, D. and Ulrich, K. (1998): Planning for product platforms, Sloan Management Review, 39 (summer 1998), pp. 19-31.

Sawhney, M. (1998): Leveraged high-variety strategies: From portfolio thinking to platform, Journal of the Academy of Marketing Science, 26 (No. 1, 1998), pp. 54-61.

Shapiro, C. and Varian, H. (1998): Information rules, Boston: Harvard Business School Press, 1998.

Shardanand, U. and Maes, P. (1995): Social information filtering: algorithms for automating word of mouth, Proceedings of the CHI 1995, ACM, 1995.

Toffler, A. (1970): Future shock, Cologny, Geneva; Orbit Publ., 1970.

Tseng, M. et al. (1997): A collaborative control system for mass customization manufacturing, CIRP Annals, 46 (No. 1, 1997), pp. 373-376.

Victor, B. and Boynton, A.C. (1998): Invented here, Boston: Harvard Business School Press, 1998.

Wigand, R., Picot, A. and Reichwald, R. (1997): Information, organization and management. Chichester, New York: Wiley, 1997.

Zerdick, A., Picot, A. and Schrape, K. (2000): E-conomics, New York: Springer, 2000.

Intermediaries for the Provision of Mass Customized Digital Goods in Electronic Commerce

Schackmann, J. & Link, H.

Summary: Generally and especially in EC, customers are confronted with a great variety and quantity of products and / or services. However, the time and effort a customer can spend on searching for his preferred products and deciding about the most preferred one based on his needs and preferences is the limiting factor. The aim of this paper is to transfer the well-known concept of mass-customization to digital products and EC. Thereby, an IT-framework will be developed, that enables intermediaries to flexibly provide personalized and mass-customized customers.

1 Introduction

With the advent of the Internet traditional segmentation approaches do not meet the special challenges of the ongoing virtualization and digitalization anymore, because they do not utilize the special possibilities of Information Technology (IT) and Electronic Commerce (EC). In this so-called information age economy segmentation approaches are superseded by IT-enabled one-to-one marketing and mass customization in order to individually target customers according to their specific needs and preferences (see e.g. [5], [11], [24], [26]; on mass information and customization systems [1], [12], [25], [34], [35]).

Generally and especially in EC, customers are confronted with a great variety and quantity of products and / or services. However, the time and effort a customer can spend on searching for his preferred products and deciding about the most preferred one based on his needs and preferences is the limiting factor. Therefore, two independent developments could be watched lately: on the one hand, new IT-enabled methods have been developed and applied to automatically match products on the one hand and the customer's interest on the other hand, both of which are described by a fixed set of attributes. Hence, a customer and product model and intelligent matching-algorithms have to be developed to satisfy the needs of customers and to provide mass customized products (See e.g. [4], [9], [14], [16],

[32]). On the other hand, the concept of an information intermediary has been introduced by several authors as a promising concept in order to establish customer (trust) relationships which are necessary means in order to get to know the customer and to deduct his needs and preferences [16], [20], [30]. Therefore, Customer Relationship Management (CRM) that enables firms to individually and professionally mass customize products has become increasingly important.

Consequently, the aim of this paper is to transfer the well-known concept of mass-customization to digital products and EC in order to evaluate the differences and specific problems and to set up a framework that enables intermediaries to provide mass-customized digital goods. The paper is organized as follows: After these introductory remarks, we will define and explain the special properties of digital products and electronic commerce in comparison to traditional products and markets in section 2. Section 3 describes the consequences of these differences for mass customization. Consequently, a framework for the mass customization of digital products will be derived in section 4. In section 5 it will be shown why intermediaries are especially suited to provide mass customized digital products. We will discuss some limitations of the model and prospects for further research in section 6. In our research we draw from the German National Science Foundation (DFG) funded theoretical research.

2 Digital Goods in Electronic Commerce

Trading with digital goods can hardly be compared to the market of "traditional goods". On the one hand, producers and distributors of digital goods might profit from new chances and possibilities provided by the nature of digital goods, whereas on the other hand they have to deal with new threats and challenges. The reason for these developments are based on the very special nature of digital goods.

2.1 The Nature of Digital Goods

The nature of digital goods is constituted in their immaterial form of bits (represented by a binary code of 0 and 1), which leads to the property of having no weight and moving with speed of light. Moreover, digital goods can be copied, causing hardly any costs, whereas the copy cannot be distinguished from the original [22]. Thus the marginal costs of one more copy of a digital good are zero. Consequently, digital goods can easily be produced and distributed on networks, such like the internet hardly causing marginal costs or time of delivery and without having to be transformed or changed in media. Examples of digital goods are *digital products* like software tools, *digital services*, such as digital information, and *digital rights*, e.g. financial products. In this paper, we will focus on the latter two, since those are mainly traded in EC.

Finally – and probably most importantly -, digital goods can easily and without costs be varied and differentiated, since they can simply be unbundled into atomic units and bundled again according to any consumers' needs [2]. All product combinations are attainable.

2.2 The Nature of Electronic Commerce

The core difference between traditional business and EC is based on the subject of trade: whereas the old economy deals with physical goods and services, electronic commerce is focused on digital goods like digital information or digital services. This leads to a new economy and a new way of doing business, where traditional and already acknowledged theories have to be critically reviewed and new rules have to be defined, new variables have to be considered. Whereas the traditional economy could mostly be restricted to a certain local area (local competition), a producer in EC automatically becomes a global player [8]. The Internet as distribution channel reaches all Internet users around the world, regardless of place and time.

According to Porter [27], a firm can have two basic types of competitive advantages: cost leadership or differentiation. Still, a firm cannot pursue both strategies, because it will be stuck in the middle. However, a producer of digital goods trying to gain advantages from a strategy of differentiation will fail, since digital goods can easily be transformed or varied. Thus, if a producer offers an innovative digital product, every other producer can imitate this good, failing to gain competitive advantage by differentiation to both of which. Pursuing a cost leadership strategy, firms have to compete in prices. This might be a fatal strategy in EC as well, since the neglectable marginal costs of production and the winner-takes-all-properties of such markets might lead prices down to zero according to microeconomic theory. Consequently, when producing digital products, a competitive advantage can neither be gained through differentiation, nor by pursuing a cost leadership strategy.

2.3 The Mass Customization Strategy in Electronic Commerce

A winning strategy in EC might be the customization of digital goods according to the individual needs and preferences of each customer. A customized product is unique, since it is perfectly adjusted to the needs of one single customer. The incentive for other firms to imitate a digital good diminishes, because the product does not satisfy the individual needs of another customer. Therefore, a competitive advantage according to Porter's theory can be achieved. Moreover, the producer might even be able to gain advantage through cost leadership, since digital information goods can easily be customized by the help of modern and innovative IT. Consequently, a producer of digital goods might gain competitive advantage through customization as well as low costs (see [25]).

The key to the success of this new strategy of customization is the knowledge of the customers' needs and preferences. In order to get to know the customers' preferences, each client has to be integrated in the production process. This integration in the producers value chain leads to the new customer status of "prosumer"[25], which is a combination of the client as *producer* as well as *consumer*. The prosumer's preferences are an important input in the production and adjust the digital good to the specific customers needs.

To sum up, the main factor leading to the success of customization lies in the customer know how. The company which will be most successful in getting detailed customers needs and moreover, will be able to build digital goods fitting these needs, will gain competitive advantages. This might be the key to the success in the new economy.

3 Mass Customization of Digital versus Physical Goods

In general, - i.e. in relation to traditional physical goods - [1], [25], [35] describe the main challenges and problems of mass customization as follows:

- A very flexible manufacturing organization and control is required, in order to produce a number of differentiated products in arbitrary order.

- The distribution and logistics has to fit and enable a such kind of flexible manufacturing process.

- The provision and maintenance of this infrastructure induce additional costs of production, which have to be justified by sufficient additional revenues.

Consequently based on the results of chapter 2, these challenges and problems do not seem to apply anymore for digital products and EC, since they can be bundled and unbundled without any costs [2]. However in EC, the depicted problem of a flexible production process is substituted by the problem of automatically integrating the customer (its attitudes, preferences, tastes etc.) as well as the products (attributes) into the customization process:

- How has the customer interface to be designed in order to receive relevant information?

- How can a customer's needs and preferences be derived from this information?

- What is the adequate form of representation of this information and know-how?

- How can a customer's needs and preferences be matched with the available assortment of products?

As it can be seen easily, the focus has switched from the product side to the customer side of the process, since product diversification is assumed to be trivial and without any costs, but the proceedings of getting to know your customer and offering him individual products seems to be the decisive competitive advantage. In the next section, it will be shown how these results will enable a new business model. Therefore, in the next section a framework will be presented, that incorporates these results.

4   Framework for Provision of Customized Digital Goods

In general - regardless of digital or physical, and customized or standardized products - the problem of providing customers with personalized products to their problems is a very complex one. Firstly, the customer himself has to be modeled by means of a machine readable representation of his (changing) preferences, attitudes and (latent) needs. Secondly, the products have to be described by their relevant product attributes. Finally, intelligent matching algorithms are needed to combine the customer on the one hand and the products on the other hand, that is, there has to be a matching based on the information provided in the customer and product models in order to get a customized product. This basic architecture is represented in figure 1 (based on [9], [16]).

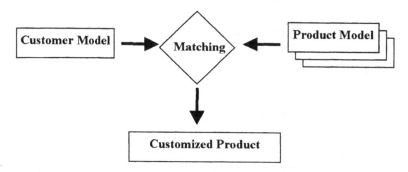

Figure 1: Basic Matching Process

4.1 Customer Model

Although the concept of customer modeling has its origin in the late seventies, only recently especially for its application in EC, there has been written a lot about customer modeling in literature (see e.g. [9], [10], [21], [28], [36]). However, so far there is a lack of customer models that combine both quantitative (such as income) and qualitative data (such as taste or the preference for certain products). Therefore, a customer model will be developed, that also represents qualitative data on a higher level of abstraction, that can be applied in various situations within one domain.

Usually information about customers is not scarce, but distributed throughout companies, and even if it was located at one central database to the customers' individual sales assistant or consultant, in order to figure out the customers' best fitting products, consultation requires not only data but information and knowledge on a higher level of abstraction. The aim of this customer model is to provide a central IT-enabled repository of data, information and knowledge about the customer that is applicable for the customization of digital products [9], [10].

- **General knowledge about the domain**

    In order to deduct a customers needs, attitudes and preferences, the possible set of relevant problems as well as the set of solutions within a domain are to be known. This so-called domain knowledge is therefore a basic necessity within a customer model.

- **Attitudes as knowledge about individual customers**

    Attitudes are considered appropriate to express a customer's basic and under-lying persistent (but not necessarily static) motives for the explanation of her behavior (for detailed discussion and definition see [9], [10]). From the atti-tudes the customer's preferences can be deducted (e.g. multi-attribute value functions, see e.g. [7]). Individual attitudes are not permanent but may change over time by a permanent update of the user model triggered by new informa-tion and data.

- **Information**

    In order for a sales assistant or a consultant to customize or even to just param-eterize products, not only knowledge, but plain information and data might be required as well.

By incorporating domain knowledge, knowledge about the customers represented by their attitudes and customer data and information, this concept of customer modeling can be applied in various domains, as well as one specific customer model can be applied for various kinds of problems and consulting situations within one domain (see figure 2).

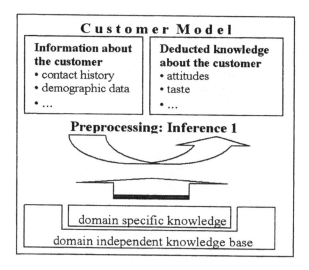

Figure 2: Customer Model

4.2 Product Model

A product model ensures that the information about the available products needed to identify the right product for a specific customer, is accessible to an automatic matching process [16], [19]. Therefore, the relevance of a product attribute is deducted from its relevance for the customers' buying decision, since it is the customer's needs which have to be satisfied with the matching process using the attributes.

The elicitation of the product attributes from the product properties can be done in several ways. Most commonly it is probably done by humans. However, an application of some kind of artificial intelligence might be feasible as well. More information about meta modeling can be found at http://dublincore.org; for various methods and applications of attribute elicitation see [31]. However, the product description with adequate attributes is a onetime process for each product. Hence – compared to the customer model -, it neither causes much effort, nor does the elicitation process seem to be very sophisticated.

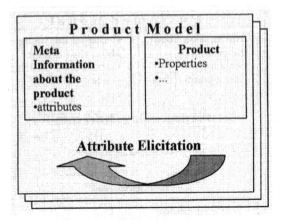

Figure 3: Product Model

4.3 Matching Inference Mechanisms

Based on the depicted customer and product model, a more sophisticated 2-step inference process can be derived. The main features of this process can be described as follows [9], [10]:

- The inference process I_1 deducts the customer's attitudes, corresponding to her needs, from the customer information base built up. This deduction is done by using domain specific and domain independent knowledge about building customer models.

- Inference process I_2 is the actual sales assistance or consulting process, which matches the customer model with the product models of the available products and thereby derives the customer's product preferences. Hence, the most preferred product will be offered to the customers. This process is supported by a domain specific and domain independent knowledge base built up for consulting processes as well. I2 refers mainly to the attitudes, but is not limited to it, e.g. for parameterization of selected product offers.

- As described in the previous section, the product attribute elicitation process is not a vital step within the mass customization process, especially since it would not be a dynamic and ongoing but onetime event. Therefore, it will not be considered part of the matching process, but a prerequisite, and therefore excluded from further considerations.

So far, there exist several approaches applying different inference mechanisms within customization systems, e.g. Broadvision (www.broadvision.com) uses a rule-based system, NetPerceptions (www.netperceptions.com) a collaborative filtering system, and Autonomy (www.autonomy.com) applies a combination of neural networks and bayesian probabilities. Moreover, other mechanisms exist, like nearest-neighbor-algorithms or ideal vector models, which are more of academic interest so far. However, there is no analytical research about the eligibility of the various mechanisms for the given problem available. See e.g. [3], [23], [29], [31] for basic information about matching algorithms; [33] discusses two matching techniques (rule base matching and collaborative filtering) for individually addressing virtual community member segments; for an economic analysis see [17], [18].

5 Intermediaries for Provision of Customized Goods

5.1 The Concept of Intermediaries

Traditionally, intermediaries are seen as institutions that by lowering transaction costs of doing business introduce a further step into the value chain [30], [37]. Thus an intermediary that provides mass customized products needs to lower transaction costs for either customers or producers.

In the context of this paper this requirement seems to be fulfilled for several reasons. With the switch of focus to the customer, each producer of customized products is expected to set up an infrastructure based on chapter 3. However, it might be very inefficient that each firm, who wants to sell to a specific customer, needs to set up this infrastructure, and as well, each firm needs to generate and store the data, information, and knowledge about this customer. Consequently, it seems rather rare that one firm possesses all necessary information, but more realistic that valuable customer information is spread all over the – usually competing - firms.

Moreover, a customer does usually not have a full, transparent market overview. Thereby, she either has to invest in market search in order to get her most preferred product or mass-customizer, or she will receive an inferior good. Especially in the context of EC, special kinds of intermediation have been discussed. [38], [30] introduced the concept of a matchmaker and a marketmaker intermediary. The first ones "acquire property on the goods traded, take over risks" whereas the latter ones "facilitate the exchange of goods and services by matching buyers and sellers without taking ownership" [38]. However with digital goods and services (compare to 2.1), the matchmaker and the marketmaker can hardly be differentiated anymore. Due to the zero costs of copy and distribution, it does not make a difference, if the intermediary becomes the owner of the good, since it will be transferred instantaneously to the customer over the Internet. Hence, in the remainder of this paper, the term matchmaker will be used.

[16], [20] introduced the idea of an intermediary that "owns" the customer trust relationship and thereby possesses all relevant data. [30] therefore introduced the more abstract concept of an information intermediary, which "is an independent profit-maximizing economic information processing system performing its activities [...] on behalf of other economic agents' information needs" [30]. Moreover, because of the amount of personal data required to customize to one customer's needs and preferences, the information intermediary is assumed not only to collect and process data, but to have a trust relationship to its customers. This enables the intermediary with the help of a relationship management to collect and aggregate the customers' data. We therefore draw the conclusion that this kind of depicted information intermediary might very well be able to reduce transaction costs and will do business more efficiently .

5.2   Discussion of Possible Business Models

Based on the modular framework (chapter 4), different business models can be derived. According to the previous discussion of intermediaries, by different combinations of the three modules of the framework (Customer model, Product model, Matching), different functional roles of the intermediaries become evident (see figure 4)[1]:

- **Traditional business model**

  In the traditional business model, which is still prevailing even in EC, no intermediaries exist. The producer of goods and services also owns the customer relationship including the relevant data and information. Hence, he is the only one able to match the customer data with its products. This model is applicable for digital as well as for physical goods. Whereas it is not only restricted to a specific domain, but its very special product assortment.

- **Customer Relationship Manager (CRM)**

  In comparison to the traditional business model, the customer relationship management model separates the producer of goods, hence product model, from an intermediary integrating customer model and matching. In this case, the producer does not have any customer data and no interface to any customers, but becomes a business-to-business player. The relationship manager owns the customer relationship as well as the customer data and integrates the product models from various producers. Thus, he is capable to match the customer

---

[1] The different business models are named according to the intermediary closest to the customer.

with the products. The CRM is not restricted to one producer, but to a certain domain due to the required domain knowledge incorporated in the customer model (see 4.1).

- **Personal Data Environment Provider (PDEP)**

  Finally, the most disintegrated approach is to separate all three modules. Thereby, the personal data environment provider collects, aggregates, retains and distributes the personal customer data, information and knowledge [6]. The match maker integrates the data of the PDEP and the producer in order to match the customer with his most preferred product. While in this model the PDEP is domain independent, again the matchmaker is restricted to a certain domain.

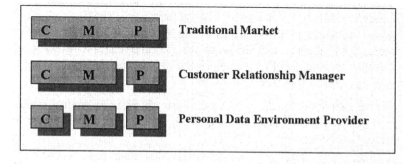

Figure 4: Functional roles of intermediaries

6 Conclusion

It has been shown that there exist significant differences between the mass customization of digital and physical products:

- Digital products can easily be unbundled to atomic units and rebundled according to a specific customer's needs and preferences with no additional costs.

- Traditional mass customization approaches focus on the product side. However, with EC and digital products, the competitive advantage is to know, which customer needs which differentiated product. Consequently, the matching process of the given product attributes with the derived customer attitudes is the challenge.

- Unlike traditional markets, the mass customization of digital products in EC is not an strategic option, but a necessity. Hence, the application of mass customization will become a competitive advantage and a focus of further work, in practice as well as in research.

Based on these results a framework has been developed, which represents this increasing customer focus and enables an automated, IT-enabled consultation process matching customer data with product data. Thereby this framework laid ground for a detailed analysis of possible business models.

Prospects for further research are:

- A powerful customer interface is required for a successful CRM, that provides the customer information and know-how for effective mass customization.

- The implementation of a behavioral model for the description and forecast of customer needs and preferences – e.g. on the basis of attitudes - in a customer model provides a powerful means for the succeeding matching process. Consequently, further research should on the one hand focus on the explanation of customer behavior, and on the other hand on the representation of thereby derived customer know-how.

- For the description of the products by means of relevant product attributes, a meta model and language, like e.g. XML, is required, that is applicable for various kinds of product categories.

- In order to efficiently match the customer model with products, a taxonomy of matching problems and adequate matching inference mechanisms is to be developed.

- Especially in EC, an high performance IT system is key for satisfied customers, customer retention and high sales. Therefore, research focus should also be on efficient IT-infrastructures [13], [15].

References

[1] Anderson, D. M.: Agile Product Development for Mass Customization; Irwin, Chicago 1997.

[2] Bakos, Y., Brynjolfsson, E.: Bundling information goods: Pricing, profits and efficiency. In Hurley, D.; Kahin, B.; Varian, H. (eds): The Economics of Digital Information Goods. MIT Press, Cambridge, Massachusetts, 1998.

[3] Bibel, W.: Wissensrepräsentation und Inferenz: Eine grundlegende Einführung; Verlag Vieweg, Braunschweig 1993.

[4] Buhl, H. U., Kundisch, D., Leinfelder, A., Steck, W.: IT-Enabled Sophistication Banking; in: Proceedings of the 8th European Conference on Information Systems ECIS 2000, Wien 2000.

[5] Buhl, H. U., Wolfersberger, P.: Neue Perspektiven im Online- und Multichannel Banking; in: Locarek-Junge, H.; Walter, B. (eds): Banken im Wandel: Direktbanken und Direct Banking, Berlin-Verlag, Berlin 2000.

[6] Davidson, Alan: Internet Everywhere; Privacy Nowhere: The Importance of Privacy by Design in the Personal Data Environment. http://www.cisp.org/imp/june_2000/06_00 davidson.htm, download 07/15/2001.

[7] Davison, M. L.: Multidimensional Scaling; Wiley & Sons, New York 1983.

[8] Dümpe, Oliver: Chancen und Risiken von Netzmärkten für Anbieter - eine spieltheoretische Analyse. Discussionpaper WI-69 of the Business School at the University of Augsburg, September 1999.

[9] Fridgen, M., Schackmann, J., Volkert, S.: Preference Based Customer Models for Electronic Banking; in: Proceedings of the 8th European Conference on Information Systems ECIS 2000, Wien 2000.

[10] Fridgen, M., Volkert, S. et al.: Kundenmodell für eCRM - Repräsentation individueller Einstellungen. Accepted paper for: 3. FAN-Tagung 2000, Siegen, Oktober 2000.

[11] Hansen, R.: Conceptual Framework and Guidelines for the Implementation of Mass Information Systems; in: Information & Management, 28th Vol. 1995, no. 3.

[12] Hansen, R., A. Scharl: Cooperative Development of Web-based Mass Information Systems. In: Proceedings of the 4th Americas Conference on Information Systems (AIS '98), Baltimore 1998.

[13] Homburg, C., Giering, A., Hentschel, F.: Der Zusammenhang zwischen Kundenzufriedenheit und Kundenbindung; in: Die Betriebswirtschaft, vol. 59, 2/1999, pp. 174-195.

[14] Kobsa, A.: Benutzermodellierung in Dialogsystemen, Springer, Berlin 1985.

[15] Krishnan, M. S., Ramaswamy, V., Meyer, M.C., Damien, P.: Customer satisfaction for financial services; Management Science, vol. 45, September 1999, pp. 1194-1209.

[16] Kundisch, D., Wolfersberger, P., Calaminus, D., Klöpfer, E.: Enabling eCCRM: Content Model and Management for Financial eServices; in: 34th Annual Hawaii International Conference on System Sciences (HICSS) 2001, Maui, 2001.

[17] Link, H., Schackmann, J.: Individuelle digitale Güter und Leistungen im Electronic Commerce. Submitted paper for WIRTSCHAFTSINFORMATIK. Discussionpaper WI-98 of the Business School at the University of Augsburg, June 2001.

[18] Link, H., Schackmann, J.: Ein ökonomisches Modell für die Produktion individueller digitaler Produkte. In: Bodendorf, F.; Grauer, G.,(eds.): Verbundtagung Wirtschaftsinformatik 2000, Siegen, Oktober 2000, Shaker, Aachen 2000, pp. 192 - 207.

[19] Lincke, D., Schmid, B.: Mediating Electronic Product Catalogs: Association for Computing Machinery; in: Communications of the ACM, New York, July 1998.

[20] McKnight, Harrison D., Chervany, Norman L.: Conceptualizing Trust: A Typology and E-Commerce Customer Relationships Model. In: Proceedings of the 34th Hawaii International Conference on System Sciences, 2001.

[21] Mertens, P., Höhl, M.: Wie lernt der Computer den Menschen kennen? Bestandsaufnahme und Experimente zur Benutzermodellierung in der Wirtschaftsinformatik; in: Scheer, A.-W.; Nüttgens, M. (eds.): Electronic Business Engineering, Physica Verlag, Heidelberg 1999.

[22] Negroponte, N.: being digital; Hodder & Stoughton, London 1995.

[23] Pau, L. F., Gianotti, C.: Economic and Financial Knowledge-Based Processing; Springer Verlag, Heidelberg 1990.

[24] Peppers, D., Rogers, M.: The one to one future; Currency Doubleday, New York 1997.

[25] Piller, F. T.: Mass Customization. Ein Wettbewerbskonzept für das Informationszeitalter; Gabler Verlag, Wiesbaden 2000.

[26] Pine II, B. J., Peppers, D., Rogers, M.: Do you want to keep your customers forever? Harvard Business Review, March/April 1995, pp. 103-114.

[27] Porter, M.: Competitve Strategy; New York 1980.

[28] Probst, A., Wenger, D.: Elektronische Kundenintegration, Marketing, Beratung & Verkauf, Support und Kommunikation; Vieweg Verlag, Braunschweig/Wiesbaden 1998.

[29] Rehkugler, H., Zimmermann, H. G. (eds.): Neuronale Netze in der Ökonomie: Grundlagen und finanzwirtschaftliche Anwendungen; Verlag Vahlen, München 1994.

[30] Rose, Frank: The Economics, Concept, and Design of Information Intermediaries. Physica-Verlag, Heidelberg, New York 1999.

[31] Schackmann, J., Knobloch, M.: Web-Mining mit Methoden des Information Retrievals - Individualisierung von Web-Sites auf Basis von Webtracking Daten. In: 5. Internationale Tagung für Wirtschaftsinformatik, Augsburg, 2001.

[32] Schneider, J., Buhl, H. U.: Simultane Optimierung der Zahlungsströme von Leasingverträgen und deren Refinanzierung; in: Zeitschrift für Betriebswirtschaft, 69, Ergänzungsheft 3, 1999, pp. 19-39.

[33] Schwaiger, M.: Virtual Communities: Der gläserne Kunde wird Realität,; in: Riekeberg, M. v.; Stenke, K. (eds): Banking 2000 - Perspektiven und Projekte, Gabler, Wiesbaden 2000, pp. 175 - 188.

[34] Wells, J. D., Fuerst, W. L.,. Choobineh, J.: Managing information technology (IT) for one-to-one customer interaction; in: Information & Management, vol. 35, 1999, pp. 53-61.

[35] Wiegran, G., Koth, H.: firma.nach.maß, Markt & Technik Verlag, 2000.

[36] Will, A.: Repräsentation finanzwirtschaftlicher Probleme für Anwendungen der verteilten künstlichen Intelligenz; Beitrag zur 6. Tagung Geld, Finanzwirtschaft, Banken und Versicherungen, Karlsruhe 1993.

[37] Williamson, Oliver.: The Economic Institutions of Capitalism. Free Press, New York, 1985.

[38] Yavas, A.: Marketmakers versus Matchmakers. In: Journal of Financial Intermediation, 2, 1992, pp. 33-58.

Knowledge Fusion in the Business Information Environment for e-Manufacturing Pursuing Mass Customisation

Smirnov, A., Pashkin, M., Chilov, N. & Levashova, T.

Summary: Modern trends in knowledge-dominated economy / society are (i) from "capital-intensive business environment" to "intelligence-intensive business environment" and (ii) from "product push" strategies to a "consumer pull" management. This leads to Mass Customisation (MC), assuming mass productions of individually customised goods and services at mass production costs, and e-Manufacturing, based on intensive use of WWW-technologies in the manufacturing part of e-business. Technology of Knowledge Fusion (KF), based on the synergistic use of knowledge from multiple sources, is a good basis for MC information support in e-manufacturing. The paper discusses a concept of a KF technology and its applications in the area of e-manufacturing based on utilizing ontologies and intelligent agents.

Keywords: mass customisation, business information environment, knowledge fusion, ontology management, intelligent agents

1 Introduction

The global business information environment in the field of Business-to-Business or virtual supply networks is the most advanced area in regard to the quantity of information being interchanged and information technologies being used, since e-business requires cooperation and open standard-based information/knowledge exchange between all the participants in real-time environment. E-Manufacturing as a part of e-business requires IT systems to react dynamically to unforeseen changes and unexpected user needs. This determines the major trends of knowledge-dominated economy: (i) shift from "capital-intensive business environment" to "intelligence-intensive business environment" – an "e" mindset – and (ii) shift from "product push" strategies to a "consumer pull" management – Mass Customisation (MC) approach. These trends as faced by all players/users in the global business information environment, including end-users/customers and knowledge-source organisations (companies, government agencies, etc.), have an increasing importance for (1) networking communication; (2) knowledge assets; (3) co-ordination of structure; (4) customer focus and real-time responsiveness to current situation trends; (5) flexibility, adaptability, agility and mobility; (6) organisational virtuality; (7) shared values and integrity; (8) ability to operate globally and in tune with local/particular problems; (9) learning and innovation.

This shift from Electronic Data Interchange to Internet-based communication systems requires global changes in business paradigms. While the traditional thinking about manufacturing involves people, process, and technology, e-manufacturing considers two more factors: infrastructure and strategy. One of the major constraints in the widespread acceptance of e-manufacturing and e-business communities is the difficulty of collective bringing together many disparate industry competitors, non-industry players, and other participants / units, and ensuring a common level of knowledge, understanding, and commitment. These communities require cooperation and open exchange of information among all participants.

The largest single driver of this is the continuing virtualisation of major corporations and increasing cooperation among smaller firms and within organisations. This virtualisation forces enterprises to find new methods of collaboration with partners, suppliers, advisors, and others, not only for the sake of knowledge transfer, but as a core integration process in development and delivery of goods and services. Virtualization causes a re-evaluation of the relationships between employee and intellectual capital. By the knowledge brokering markets and expertise profiling in web-based collaboration tools, teams and companies can temporarily seek talent from non-employees via the Internet. The adaptability and flexibility of an organisation depend not only on the electronic relationships between partners, suppliers and customers, but on the organisation's ability to quickly react to market, technology and relationship changes.

Thereby MC approach requires companies to intensively collaborate, and flexibility requires relationships to be loose same time. Virtual Enterprises (VE) seem to be a logical solution for this (Rautenstrauch and Turowski, 2001) since it is a temporal cooperation of independent units providing a service based on the shared products, technologies (and processes), and resources for changing market needs and built on a common business understanding. Also they suggested VE model for MC consisting of MC-macro-processes (which includes design, manufacturing and distribution processes), IS support, flexible coordination and role concepts. Outsourcing is the core of VE, when organisations rely on third parties in both products production and process technology use. Development of information technologies also made possible many companies to outsource their IT function (Caddy, 2000). In this case IT support can be seen as an extension of IS support including not only systems but the entire information environment. Hence, in the case of IT function outsourcing by the third party companies, these companies can be considered members of VE with IT function becoming a part of the MC-macro processes (Figure 1). As a result IT function is getting strongly involved into manufacturing process.

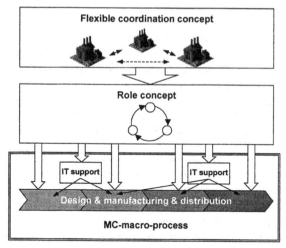

Figure 1: VE model for MC (adapted from Rautenstrauch, Turowski, 2001)

Among the options for maximizing a value from supply chains are two major strategies: (i) *postponement* (delayed differentiation until customer's demand for specific end products) and (ii) *information sharing* (for faster and more accurate information flow across the supply chain). A *postponement* is a special form of MC. It assumes withholding any modification or customisation of the product (keeping it generic) as long as possible (Lee and Billington, 1995). The principle of postponement can be stated as: "The time of shipment and the location of final product processing in the distribution of a product should be delayed until a customer order is received" (Zinn and Bowersox, 1988). Certain business conditions that favour each strategy are depicted Table 1 (Billington and Amaral, 2000).

The integrated approach (postponement & information sharing) for "product-process-resource" system configuration management is a new way to make build-to-order in e-business and e-manufacturing for postponement & responsiveness. This approach could maximize profitability by capturing the value of postponement.

Table 1: Business favouring major strategies for MC

Strategy	Demand Uncertainty	Capacity Re-sponsiveness	Cost of a Lost Sale	Cost of Post-ponement
Postponement	High	Low	High	Low
Information	Low	High	Low	High

Sharing				

All changes required for implementation of MC approach are connected to changes in information factors: production system needs more knowledge about customers and customers need more knowledge about products (Caddy, 2000). This means that information can be regarded as the most important factor for the implementation of MC (Reichwald, Piller, and Moeslein, 2001). "Being truly customer focused is not possible if the organisation is not, first, information intensive" (Blattberg and Glaser, 1994). Customer Relationship Management includes processing, storing, updating and analysing customer specific information along with each individual order to be controlled (Seelmann-Eggebert, 1999).

Knowledge, as a development of information, can be defined as a set of relations (constraints, functions, rules) by which a user/an expert decides how, why, where and what to do with the information to produce timely adaptive actions meeting a goal or a set of goals. Now, it has become a critical resource for any activity more powerful than natural resources. Capitalising on collective knowledge and intellectual assets takes more than Internet/Intranet. In order to capitalise on this knowledge organisation/end user has to organise and manage it in a creative and useful way. As a result, new information technologies, such as product data management and knowledge management are increasingly used based on ontology management and multi-agent technologies (Fikes and Farquhar, 1997; Fischer et al., 1996; Jennings, 2000; Noy and Musen, 1999; Smirnov and Chandra, 2000; Wooldridge and Jennings, 1995).

As a result, there arises a need to develop a cost-effective approach to integration of knowledge from distributed sources for global awareness, dynamic planning and global information exchange.

2 Knowledge Fusion Concept

The approach of Knowledge Fusion (KF) implies integration of knowledge from different sources (probably, heterogeneous) into a combined resource in order to complement insufficient knowledge and obtain new knowledge. Thereby it is based on the synergistic use of knowledge from multiple sources. KF is being integrated into an information environment for two principal areas: (i) customer KF supports traditional activities with on-line sources and uses a customer profiles, and (ii) KF portals support shared knowledge, an inventory management of the internal and external knowledge sources. The environment includes end-users/customers, loosely coupled knowledge sources/resources, and set of tools and methods for information processing (Smirnov, 2000). The information environment is called "scalable" if it can be scalable against both the number of its members and the number of various types of its members (Smirnov, 2001b). In

(Smirnov, 2001a) characteristics and requirements for such environment, particularly for business information environment, are described.

Analysis of some existing systems/projects for knowledge source integration is presented in Table 2 (Aguirre, Brena, and Cantu, 2001; Gray, Hui, and Preece, 1999; Jacobs and Shea, 1996; Preece et al., 2000). These systems are listed below:

- KRAFT (Universities of Aberdeen, Cardiff and Liverpool in collaboration with British Telecommunications PLC) – Knowledge Reuse and Fusion / Transformation – multi-agent system for integration of heterogeneous information systems. The main aim of this project is to enable sharing and reuse of constraints embedded in heterogeneous databases and knowledge systems.

- InfoSleuth (Microelectronics and Computer Technology Corporation (MCC), Austin, Texas) – multi-agent system for retrieving and processing information in a network of heterogeneous information sources.

- OBSERVER (University of Zaragoza, Spain) – Ontology Based System Enhanced with Relationship for Vocabulary heterogeneity Resolution – system for information retrieving from repositories. The main aim is to retrieve information from heterogeneous knowledge sources without having knowledge of their structure, location and existence of the requested information.

Having analysed existing systems the following advanced requirements for knowledge-based systems could be identified:

- *Velocity.* The advanced organisation constantly seeks for the ways to reduce and/or compensate for a variability in a customer/user demand and suppliers/sources. This could involve negotiations with multiple suppliers, optimising the temporary work force, and eliminating inefficiencies in the information processes.

- *Flexibility.* The organisation leaders must be ready for sudden changes in target problem demand. The organisation maintains its flexibility by keeping minimal information volume in the sources.

- *Integrity.* Monitoring of system's components for their availability and changes.

Table 2: Some information integration systems

Feature	KRAFT	OBSERVER	InfoSleuth
Project platforms	KQML P/FDM, CoLan	Description Logics (DL) CLASSIC	At first - KQML, KIF; ODBC; at present – OKBC, JDBC. LISP, CLISP, LDL+, Java, C/C++, NetScape
Knowledge / information sources	Any available sources for which special processing tools developed	Relational and object – oriented databases, HTML pages.	Initially – databases; currently – any available sources
Basic ontologies	WordNet	Depends on application area	Depends on application area
User can choose ontology to work with	No	Yes	Yes
Ontology relationship	Hierarchy	Lattice	Meta Level Ontology for other ontologies translation
Ontology extending	Yes	Yes	Yes
Ontology-specification language	P/FDM	DL	Initially - KIF and LDL+; currently – OKBC
Agent-based architecture	Yes	No	Yes
Agent coordinator	Facilitator		Broker Agent
Main types of agents	Wrappers (interaction with sources), Facilitators ("yellow pages" directory service for the agents), Mediators (task execution control), User Agent (interaction with user)		Resource Agent (Wrapper in KRAFT), Broker Agent (Facilitator in KRAFT), TaskExecution Agent (Mediator in KRAFT), User Agent

- *Open Connectivity*. The organisation's knowledge should be available for the shared access by external users.

- *Reasoning*. The organisation must have reasoning for proposed solutions and clear plan of actions to achieve its goals.

– *Customisability.* The organisation must be ready to build any required configuration of knowledge domain model. Besides, it must be able to motivate the suggested solution.

– *Learning from the user.* In case of some external impacts or actions it is necessary to provide for an ability to include required changes into the system behaviour rules (Ramos, 2001).

Ontologies were chosen as a common model of knowledge representation for KF operations. The generic scheme, according to which KF could be performed, can be represented by the following scheme of ontology operations (Table 3).

Table 3: Correspondence between KF operations and operations on ontologies

Knowledge Fusion Operations	Operations on Ontologies
Acquisition	
Identifying appropriate knowledge from external sources by locating, accessing, valuing and/or filtering	Finding
Capturing identified knowledge of sufficient reliability, relevance, and importance by extracting, collecting, and/or gathering	Term translation Query Extraction
Organizing captured knowledge by distilling, refining, orienting, interpreting, packaging, assembling, and/or transforming it into usable representations	Matching & translation Slicing or Pruning
Transferring organized knowledge	
Selection	
Identifying appropriate knowledge from system's existing sources by locating, accessing, valuing and/or filtering	Finding
Capturing identified knowledge of sufficient reliability, relevance, and importance by extracting, collecting, and/or gathering	Query Extraction
Organizing captured knowledge by distilling, refining, orienting, interpreting, packaging, assembling, and/or transforming it into usable representations	Slicing or Pruning

Table 3: Correspondence between KF operations and operations on ontologies (continued)

Knowledge Fusion Operations	Operations on Ontologies
Transferring organized knowledge	
Generation	

Producing knowledge based on existing knowledge by creating, synthesizing, analysing, and constructing	Creating
	Merging:
	  Modularisation
	  Intersection
	  Union
	  Slicing or Pruning
	  Saving
	  Validation

Transferring the produced knowledge for externalisation and/or internalisation	
Internalisation	
Assessing with requisite cleaning, refining, and filtering and targeting	Pruning
Delivering the knowledge representations to targeted resources	Extension Update
Externalisation	
Embedding knowledge into system's output for release into the environment	Browsing Translation Publishing

Within the research being done the following KF patterns are identified, which could considerably accelerate the processes of KF due to typification of fusion schemes. Knowledge Source Fusion patterns are illustrated via the following example. The two sources (A and B) with some structures of primary knowledge units are given (Figure 2). There is a tacit relation between two primary knowledge units, namely a3 from A and b2 from B. It is necessary to fuse the two sources preserving the internal knowledge structure and revealing the tacit relation mentioned above.

- *Selective Fusion.* The new knowledge source is created, which contains required parts of the initial knowledge sources. The initial knowledge sources preserve their internal structures and autonomy.

– *Simple Fusion.* The new knowledge source is created, which contains initial knowledge sources. The initial knowledge sources preserve their internal structures though lose (partially or completely) their autonomy.

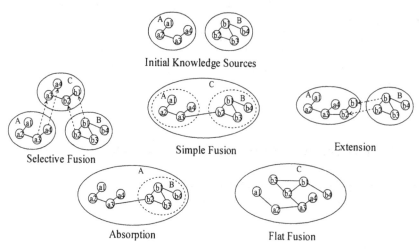

Figure 2: Knowledge fusion patterns

– *Extension.* One of the initial knowledge sources is extended so that it includes the required part of the other initial knowledge source, which preserves its internal structure and autonomy.

– *Absorption.* One of the initial knowledge sources is extended so that it absorbs the other initial knowledge source, which preserves its internal structure though loses (partially or completely) its autonomy.

– *Flat Fusion.* The new knowledge source is created, which contains initial knowledge sources. The initial knowledge sources dissolve within the new knowledge source and do not preserve their internal structures and autonomy.

Major functions of KF could be determined as (i) communication, (ii) co-ordination, (iii) collaboration, and (iv) common / shared memory. This set of functions would be realised via the following technologies (Smirnov 2000):

– direct knowledge entry by domain experts (based on GUI, complex object cloning, and object-oriented template library),

– knowledge repository parallel development by distributed teams (based on automatic change propagation and conflict resolution),

– knowledge sharing by knowledge maps (based on reusable ontology theory and distributed constraint satisfaction technology), and

- distributed uncertain knowledge management (based on object-oriented fuzzy dynamic constraint networks as a shared ontology paradigm).

The multi-agent approach is much more suited for scalability of global information environment than the conventional approach due to its, (a) orientation towards object-oriented modelling for encapsulation, and (b) suitability for unstructured knowledge problem domains, and is proposed as a next generation model for complex, distributed systems. Therefore, a multi-agent approach has been adopted for this research. The set of techniques listed above would be realized as a group of agents.

3   Multi-Agent Architecture for Knowledge Fusion

The following three business information environment components can be selected, for which agent types are to be defined:

1. Users, knowledge consumers;

2. KF system;

3. Knowledge sources.

FIPA Reference Model (FIPA, 1997-1999) has been chosen as a technological infrastructure for definition of agents' properties and functions. *User Agent* is proposed to work with a user, knowledge consumer. According to FIPA this is an agent interacting with a human being (FIPA, 1998). Knowledge sources can be divided into two groups: (i) *implicit impact sources* (knowledge bases, databases, repositories, etc.), which influence upon knowledge stored in the system indirectly and (ii) *explicit impact sources*, which can make changes to the system knowledge directly (experts, knowledge management tools). For interacting with implicit impact sources it is proposed to use a *Wrapper Agent,* providing the FIPA-WRAPPER service for an agent domain in the Internet (FIPA, 1998). This type of an agent was used in KRAFT system and InfoSleuth system (as Resource Agent). *Expert Assistant Agent* is proposed for interacting with experts.

According to FIPA it is necessary to develop the following technological agents:

- *Facilitator* – Directory Facilitator – is an agent that provides a "yellow pages" directory service for the agents. It stores descriptions of the agents and the services they offer (FIPA, 1998). Agents of this type were used in KRAFT and InfoSleuth (Broker Agent).

- *Mediator* – Information Broker Agent, which offers a set of communication facilitation services to other agents using some knowledge about the requirements and capabilities of those agents (FIPA, 1999). Mediator is proposed for task execution tracking inside the system. Similar agents were used in KRAFT and InfoSleuth (TaskExecution Agent).

Also, KF technologies require several additional agents which can be defined as follows:

- *Translation Agent* and *KF Agent* provide operation performance for KF.

- *Configuration Agent* supports effective use of knowledge source network.

- *Ontology Management Agent* provides ontology operation performance.

- *Monitoring Agent* reduces system failure probability by knowledge source verifications. Life cycle of KF systems consists of forming problem domain (preparation phase) and utilizing it with possible modification (operation phase). During the operation stage KF systems work in real-time mode. Accessibility and consistency of knowledge sources are the critical factors for them.

Depending on the problem domain specifics most of existing multi-agent systems have their own set of agents with special functions, language and communication rules. Table 4 describes some specific features of the proposed agents.

Table 4: Knowledge fusion system agents features

Agent	Life time	Quantity	General tasks
Wrapper	Knowledge source life time	Number of knowledge source types	Translates knowledge from source format representation into system format representation and sends requests from system to source.
Mediator	Task execution time	Number of tasks being processed	Tracks out task processing step-by-step from input to result. Provides *negotiations* with other agents. Stores temporary results.
Facilitator	System life time	1	Provides a "yellow pages" directory service for the agents.
User Agent	As long as user is registered in the system	Number of registered users	Provides a set of functions for *user profile* processing. Facilitates request input, provides for a set of tips and hints for user, passes messages and information from the system to the user.
Translation Agent	System life time	Number of request input interfaces.	Provides translation functions between user and system. Works with *templates*, and *application ontology*.

Table 4: Knowledge fusion system agents features (continued)

Agent	Life time	Quantity	General tasks
Expert Assistant Agent	As long as expert is registered in the system	Number of registered experts	Facilitates the process of expert knowledge input into the system. Forms expert profile.
Configuration Agent	System life time	1	Configures KSFTree[1] from different sources using *knowledge map*. Performs scheduling functions. *Negotiates* with KF Agents, Wrappers and Expert Assistant Agents.
KF Agent	System life time	Depends on problem domain	Obtains knowledge from *Mediator* and fuses it. Generates new knowledge. Validates it. Can perform some operations concurrently using different *KF patterns*.
Monitoring Agent	System life time	1	Provides a set of functions for diagnostics of internal system information base and external knowledge sources.
Ontology Management Agent	System life time	Depends on problem domain	Provides a set of functions for ontology engineering and operation – creation of ontologies for new knowledge sources, modification of application ontology etc. Checks correspondence between knowledge source and request ontologies and application ontology.

Agent structure could contain the following modules (Guan, Zhu, and Ko, 2000): (i) identifying, (ii) functional, and (iii) knowledge repository. Identifying module contains such parameters as unique identifier, creation date and time, type. Func-

---

[1] Knowledge Source Fusion Tree. The system performs decomposition of the request into its components – subrequests and sends them to processing. During the process of identifying knowledge sources for the subrequests a KSFTree structure is defined. The root of this tree is the user request, the leafs are the knowledge sources and the nodes are the subrequests obtained during decomposition of the user request.

tional module contains a set of procedures to be executed by the agent. Knowledge repository contains special information, such as a history of the agent contacts, temporary results, new knowledge, etc. Agent actions can be described as follows: if certain condition is met, then it is (obliged / permitted / forbidden) to perform a (<action / speech act>) (Barjis and Chong, 2000).

A major set of agents is represented in Figure 3 according to the above described principles and functions of the KF system. During creation of multi-agent environment, it is necessary to solve the problems related to system's effectiveness. Namely:

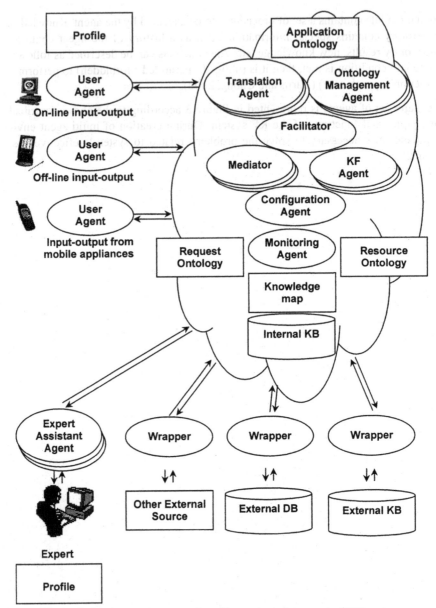

Figure 3: Basic components of multi-agent environment of KF system

– To avoid agent overloading it is reasonable to store some functions on a server using them as services for agents (Saba and Santos, 2001).

– As it can be seen, agent interaction is broad enough, therefore, it is necessary to provide solutions for such problems as estimation of throughput, creation of central work directory, creation of agent search (location) table (Saba and Santos, 2001).

– On the one hand actions of some agents limit and on the other hand they are limited by actions of other agents (Petersen and Gruninger, 2000). For instance, if one of agents works with a knowledge source using a low-bandwidth connection other agents should not use the same source.

– Since the system works in a real-time environment it is necessary to protect it from possible failures in working with sources, such as source unavailability, too long request processing, incorrect reply of the source (Goyal and Parameswaran, 2000).

4 Case Diagrams of Major System Scenarios

The procedure of the user request processing and procedure of new knowledge input by the expert for demonstration of cooperation between agents will be described below. The generic scheme of the user request processing in KF operation stage is represented in 4.

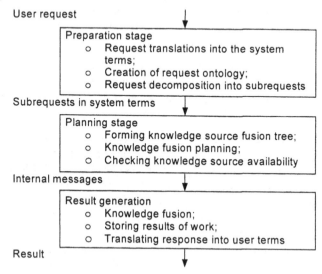

Figure 4: Generic scheme of user request processing

Case diagram of user request processing is shown in Figure 5 and represents the following scenario: (1) Mediator receives request from User Agent; (2) Mediator passes the request to Translation Agent for processing, Translation Agent and Ontology Management Agent (2.1) check it for possibility of translation into system's terms, translate it, create new request ontology; (3) Translation Agent returns translated request to Mediator; (4) Mediator passes the request received from Translation Agent to Configuration Agent. Configuration Agent performs decomposition of the request into subrequests, identifies appropriate knowledge sources, configures KSFTree and plans KF by negotiating price, time, capabilities, etc. (4.1) with KF Agents and Wrappers; (5) Configuration Agent passes results of KSFTree configuration to Mediator; (6) Mediator passes subrequests to Wrappers and Expert Assistant Agents according to the KSFTree, performs control for error, malfunction or failure occurrence; (7) Wrappers translate the request into terms of knowledge sources, pass the request to the knowledge source, receive response from the knowledge source, translate response into system's terms, returns result to Mediator. Expert Assistant Agents perform actions of narrow-specialized Wrapper, facilitate the process of expert knowledge input into the system; (8) Mediator passes results received from Wrappers to KF Agent. KF Agent performs fusion of received knowledge, validates new knowledge and checks it for relevance to the request, passes new knowledge to Monitoring Agent (8.1) for internal processing. (9) Mediator receives and stores temporary results of KF. Steps (8) and (9) are repeated several times according to KSFTree structure. (10) Mediator passes result of KF to Translation Agent for processing, Translation Agent and Ontology Management Agent (10.1) translate system's result into the user understandable form; (11) Translation Agent passes the results to Mediator. (12) Mediator passes the results to User Agent. User Agent returns the results to the user, updates user's profile. Monitoring Agent performs necessary changes in internal knowledge base and knowledge map.

Mediator used for the request processing is cloned when the request comes into the system and exists until the request processing is finished. The number of Mediators simultaneously existing in the system is only limited by hardware resources.

Case diagram of direct input of new knowledge to the system by the problem domain expert is shown in Figure 6. It represents the following scenario: (1) the expert inputs new knowledge using Expert Assistant Agent; (2) Mediator receives necessary information, passes it to Translation Agent; Translation Agent and Ontology Agent (2.1) translate it into system terms; (3) Translation Agent returns the result to Mediator; (4) Monitoring Agent catches received knowledge, checks it for consistency and originality. After successful verification new knowledge is added to internal knowledge base and knowledge map is changed.

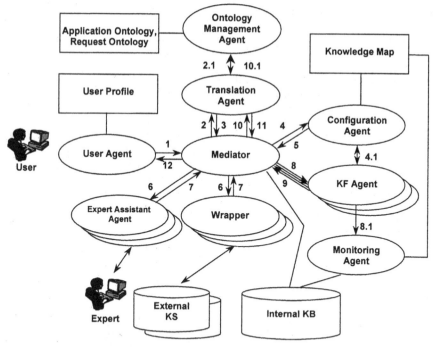

Figure 5: Case diagram of user request processing

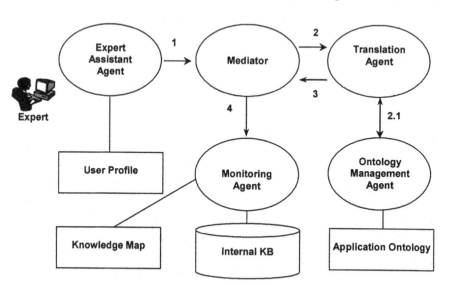

Figure 6: Case diagram of direct input of new knowledge to the system by problem domain expert

Legend for the case diagrams (Figure 5, Figure 6):

- Lines represent connections between internal information base components and agents.
- Unidirectional arrows marked with numbers represent sequential steps of scenario processing by Mediator and the steps numbers.
- Arrows marked with numbers containing points represent auxiliary operations of scenario processing without mediator. Agents do not wait for a response when arrow is unidirectional and wait for a response when it is bi-directional.
- Bi-directional arrows without numbers and text represent work with external knowledge sources.
- Facilitator is not presented in case-diagrams but it provides a "yellow pages" directory service for the agents.

5   Conclusion

In the face of globalisation in the business and worldwide increasing competition, the producers and their suppliers/dealers/partners recognise the need to achieve adequate profitability to invest in new technologies and innovations to maintain world-class position. They recognise the mutual benefit derived from a close partnership between a producer and its virtual enterprise surroundings.

The technology of Knowledge Fusion may be very useful for enabling collaboration between virtual enterprise members. Ontology-oriented structure based on widespread standards allows the implementation of effective operations with knowledge and open system architecture. Utilizing intelligent agents also increases system efficiency and interoperability.

Acknowledgements

The paper is due to the research carried out as a part of the Partner Project # 1993P according to the Agreement between The International Science and Technology Center (ISTC), SPIIRAS, and European Office of Aerospace Research and Development (EOARD), and as a part of the Project # 4.4 of the research program # 18 of the Presidium of the Russian Academy of Sciences.

References

Aguirre, J.L., R. Brena, and F.J. Cantu (2001): Multiagent-Based Knowledge Networks. Expert Systems with Applications, 2001. V. 20, pp. 65—75.

Barjis, J. and S. Chong (2000): Agent-Based E-Commerce Systems Based on the DEMO Transaction Concept and Semiotic Approach. Proceedings of the International ICSC Congress on Intelligent Systems and Applications (ISA'2000), Wollongong, Australia, pp. 474—480. ISBN 3-906454-24-X.

Billington, C. and J. Amaral (2000): Investing in Product Design to Maximize profitability Through Postponement. http://billington.ascet.com.

Blattberg, R. and R. Glaser (1994): Marketing in the Information Revolution. In: Marketing Information Revolution (eds. by Blattberg R. et al.). Boston, pp. 9—29.

Caddy, I. (2000): Moving from Mass Production to Mass Customization: the Impact on Integrated Supply Chains. Proceedings of the Workshop on Mass Customization Management (MCM 2000), University of Wollongong, Australia. (Electronic Proceedings).

Fikes, R. and A. Farquhar (1997): Large-Scale Repositories of Highly Expressive Reusable Knowledge. Technical Report, Knowledge Systems Laboratory, Stanford University, April, KSL-97-02, http://ksl-web.stanford.edu/KSL_Abstracts/KSL-97-02.html.

FIPA Documentation (1997—1999): Geneva, Switzerland, Foundation for Intelligent Physical Agents (FIPA), http://aot.ce.unipr.it/external/documentation/fipa/.

FIPA 98 Specification (1998): Part 12 - Ontology Service. Geneva, Switzerland, Foundation for Intelligent Physical Agents (FIPA).. Version 1.0. http://www.fipa.org

FIPA Spec 2 – 1999 (1999) : Agent Communication Language. Geneva, Switzerland, Foundation for Intelligent Physical Agents (FIPA). Draft, Version 0.1. http://aot.ce.unipr.it/external/documentation/fipa/.

Fischer, K., J.P. Müller., H. Heimig, and A.-W. Scheer (1996): Intelligent Agents in Virtual Enterprises. Proceedings of the First International Conference and Exhibition on the Practical Application of Intelligent Agents and Multi-Agent Technology. UK, London: The Westminister Central Hall, pp. 205—223.

Goyal, M. and N. Parameswaran (2000): Attitudes in Team Problem Solving. Proceedings of the ICSC Symposia on Intelligent Systems & Applications (ISA'2000). Wollongong, Australia (Electronic Proceedings).

Gray, P..M.D., K. Hui, and A.D. Preece (1999): Finding and Moving Constraints in Cyberspace. AAAI-99 Spring Symposium on Intelligent Agents in Cyberspace. Stanford University, California, USA. AAAI Press, pp. 121—127.

Guan, S., F. Zhu, and C.C. Ko (2000): Agent Fabrication and Authorization in Agent-Based Electronic Commerce. Proceedings the International ICSC Symposium on Multi-Agents and Mobile Agents in Virtual Organizations and E-Commerce (MAMA'2000), Wollongong, Australia. (Electronic Proceedings).

Jacobs, N. and R. Shea (1996): The Role of Java in InfoSleuth: Agent-Based Exploitation of Heterogeneous Information Resources. Technical Report, Microelectronics and Computer Technology Corporation.

Jennings, N.R. (2000): On Agent-based Software Engineering. Artificial Intelligence. N 117, pp. 277-296.

Lee, H.L. and C. Billington (1995): The Evolution of Supply Chain Management Models and Practice at Hewlett-Packard. Interfaces, Vol. 25, N 5, pp. 42-63.

Noy, N.F. and M.A. Musen (1999): SMART: Automated Support for Ontology Merging and Alignment. Proceeding of the Twelfth Workshop on Knowledge Acquisition, Modeling and Management (KAW'99). Banff, Canada. http://sern.ucalgary.ca/KSI/KAW/KAW99/papers.html.

Petersen, S.A. and M. Gruninger (2000): An Agent-based Model to Support the Formation of Virtual Enterprises. Proceedings of the ICSC Symposia on Intelligent Systems & Applications (ISA'2000). Wollongong, Australia. (Electronic Proceedings).

Preece, F., Hui, K., Gray, A., Marti, P., Bench-Capon, T., Jones D., Cui, Z. (2000): The KRAFT Architecture for Knowledge Fusion and Transformation. Knowledge-Based Systems. V. 13, pp. 113–120.

Ramos, C. (2001): Scalable Intelligence: a New Concept for the Evolution of Decision Support Systems to Intelligent System/Agents. Proceedings of the International NAISO Congress on Information Science Innovations (ISI'2001), Symposium on Intelligent Automated Manufacturing (IAM'2001), Dubai, U.A.E. (Electronic Proceedings).

Rautenstrauch, C. and K. Turowski (2001): Manufacturing Planing and Control Content Management in Virtual Enterprises Pursuing Mass Customization. Proceedings of the International NAISO Congress on Information Science Innovations (ISI'2001), Symposium on Intelligent Automated Manufacturing (IAM'2001). Dubai, U.A.E. (Electronic Proceedings).

Reichwald, R., F.T. Piller, K. Moeslein (2001): Four Strategies to Create Competitive Advantage with Customized Goods and Services on the Internet. Proceedings of the International NAISO Congress on Information Science Innovations (ISI'2001), Symposium on Intelligent Automated Manufacturing (IAM'2001). Dubai, U.A.E. (Electronic Proceedings).

Saba, G.M. and E. Santos, Jr. (2001): The Multi-Agent Distributed Goal Satisfaction System. http://www.engr.uconn.edu/~steve/Cse298300/Fall00Projs/AgentSecurity/agentsecurityspec.pdf.

Seelmann-Eggebert, R. (1999): Mass Customization: Logistik und Prozesse. Kundenindividuelle Massenproduktion – Produkte, Fertigung und Marketing. Congres Centre Festung Marienberg, Wuerzburg.

Smirnov, A. (1999a): Virtual Enterprise Configuration Management. Proceedings of the 14th IFAC World Congress (IFAC'99), Beijing, China, Pergamon Press, vol. A, July 1999, pp. 337-342.

Smirnov, A. and C. Chandra (2000): Ontology-based Knowledge Management for Co-operative Supply Chain Configuration. Proceedings of the 2000 AAAI Spring Symposium "Bringing knowledge to business Processes", March 20-22, Stanford, California, AAAI Press, pp.85-92.

Smirnov, A.V. (2000): Rapid Knowledge Mass Customization Management Infosphere: Major Requirements and Technologies. Proceedings of 2000 Advanced Summer Institute (ASI 2000) and 2000 IIMB Workshop on Integration in Manufacturing and Beyond (IIMB 2000). Bordeaux, France, pp. 356—358:

Smirnov, A. (2001a): Knowledge Fusion in the Business Information Environment: a Concept and Major Technologies. Proceedings of the 7th International Conference on Concurrent Enterprising "Engineering the Knowledge Economy through Co-operation". Bremen, Germany, pp. 273—276.

Smirnov, A.V. (2001b): Rapid Knowledge Fusion into the Scalable Infosphere: A Concept and Possible Manufacturing Applications. Proceedings of the International NAISO Congress on Information Science Innovations (ISI'2001), Symposium on Intelligent Automated Manufacturing (IAM'2001). Dubai, U.A.E. (Electronic Proceedings).

Wooldridge, M.J. and N.R. Jennings (1995): Agent Theories, Architecture, and Languages: A survey. In: Intelligent Agents: Proceedings of the Workshop on Agents Theories, Architecture, and languages (ECAI-94), Springer-Verlag, pp. 1—39.

Zinn, W. and D. Bowersox (1988): Planning Physical Distribution with the Principle of Postponement. Journal of Business Logistics. Vol. 9, N 2, pp. 117-136.

Experiences of Applying Systematic Modularisation Methods in Light Assembly Industry

Österholm, J., Tuokko, R. & Uuttu O.

Summary: Modularisation is the decomposition of a product into building blocks with specified interfaces. Today many manufacturing companies have faced challenges on an increasingly complex, customer value driven marketplace and a progressively greater number of product variants. A modular product platform is a key to handle these challenges while reducing costs and time to market. The discussed four case studies show how a modular product platform can be achieved by using a systematic working method. The method used to develop a modular product platform is also shortly reviewed in this paper. In this method so called module drivers, the reasons for dividing products into modules, play an important role in several phases of the working process. Experiences from this work with different types of products are presented in this paper. The systematic way of working eliminates market related risks, forces the project team to think in terms of product variants and future changes and to use cross-functional know-how and concentrate on customer demands. The output is a well defined product platform from which all the product variants required can be manufactured.

1   Introduction

Today's markets are faced by rapid, uncertain, and continuous change. New markets and products constantly appear, change and disappear within shorter period of time. Stiffer competition forces companies to concentrate on more and more specific market segments. This leads to increased number of product variants and to a challenge to handle a more complex product structure. Modularisation, decomposition of a product into building blocks (modules) with specified interfaces, driven by company-specific reasons, module drivers, is a way to face these challenges.

The Modular Function Deployment (MFD) method [Erixon 1998] is a systematic working method for developing a modular product platform. The method consists of five main steps. It starts with a Quality Function Deployment (QFD) analysis to clarify customer requirements and to identify important design requirements. In step two, the functional requirements on the product are analysed and the technical solutions are selected. This is followed by a systematic selection of modular concepts and the possible modules are identified with the help of the Module Indication Matrix (MIM) in step three. In step four, an evaluation can be carried out for each modular concept. Finally, in step five, the Module Indication Matrix is once again used to identify the opportunities of Design for Assembly (DFA) for further improvements. The five main steps of the MFD method are presented in Figure 1.

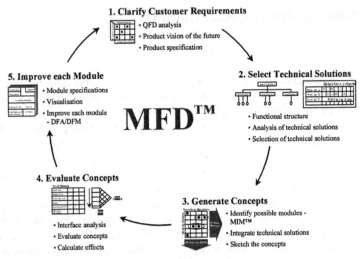

Figure 1: The five main steps of the Modular Function Deployment method [Yxkull 1999]

In reality the design work very seldom starts from the first step of the method, continues through every single step in the right order, and ends with the final step. Starting points may vary and many iterations may be needed before the end is reached. Some steps can also be carried out as parallel activities, especially steps one and two. However it is essential that, in the end, all the necessary steps have been carried out in order to reach an acceptable result.

Four case studies presented later in this paper introduce how the working method was used in product development projects in light assembly industry. In Figure 2 the case products of those four case studies as well as the main steps of the used method are presented. Those case products were (from left to right and from up to down in figure 2): digital valve controller (case 1), electrical faucet (case 2), air valve (case 3) and High Speed Assembly Cell (case 4).

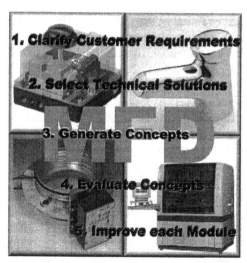

Figure 2: Four case products and the main steps of the MFD method

The central part of the MFD method is the third main step where the modular concept is generated by using a matrix called Module Indication Matrix (MIM). In the MIM, the technical solutions of the product are characterised by their module drivers which support the grouping of technical solutions into a modular concept. The module drivers are the strategic reasons for dividing products into modules. These twelve generic driving forces for modularisation that cover the entire life cycle of a product and may be linked to different functions of a company are presented in Figure 3, but they can be also complemented by company specific reasons such as: strategy, financial limitations, legal restrictions, etc.

Product develop-ment and design	– *Carry-over*
	– *Technical evolution (external)*
	– *Planned design changes (internal)*
Variance	– *Technical specification*
	– Styling
Production	– *Common unit*
	– Process / organisation
Quality	– Separate testing
Purchasing	– Supplier available
After sales	– *Service / maintenance*
	– *Upgrading*
	– *Recycling*

Figure 3: The twelve generic module drivers [Stake 1999]

A *carry-over* is a part of a product that will be used in a future product generation. It may also be carried over from one product family to another. *Technical evolution (external)* refers to a part that is likely to go through a technology shift during the life cycle of the product family. The impact here is often external. *Planned design changes (internal)* occur when a unit is estimated to go through some changes according to an internally decided plan. When the unit varies in terms of function or performance in the product family, a module driver *technical specification* is affected. *Styling* means that a part of a product varies in terms of colour and/or shape between the products in the family. An ideal situation is that these influences in trends and fashion can be limited to some modules only. Those parts and sub-functions that can be used across the whole product family are classified as *common units*. *Process / organisation* means that the unit suits a special process or has a suitable work content for a group. In some cases there is a need for *separate testing* for a part or a module before it is supplied to the main flow and that may be a reason for some part to establish a module. Modularity also provides the option to purchase complete standard modules *(supplier available)* from vendors instead of individual parts. Sometimes a quick *service / maintenance* is needed so that a damaged module can be replaced by a new one. For a company *upgrading* (the unit might be used in place of another part with a different function or performance) might gain great market advantages. *Recycling* issues concerning the end of product life-cycle are becoming and will continue to be more and more important in the future, and product manufacturers will be forced to take responsibility for recycling aspects.

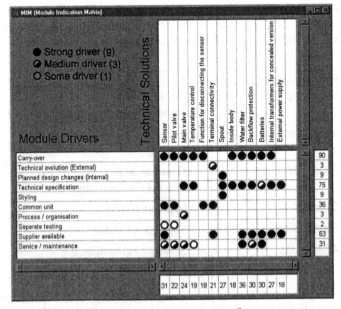

Figure 4: Module Indication Matrix (MIM) [Österholm 2000]

The Module Indication Matrix (MIM) is a QFD-like matrix in which the technical solutions of the product are assessed against the module drivers. In the MIM every technical solution is weighted on a scale: 9 (= strong driver), 3 (= medium driver) and 1 (= some driver). If there is no interaction between a module driver and a technical solution, the box is left empty (see Figure 4). This testing of technical solutions, one by one against every module driver, will form a picture of which technical solutions have many, or few, reasons for forming a module and which have the strongest reasons for becoming one.

2   Case 1 – Digital Valve Controller

In Case 1 the case product was the next generation of an existing digital valve controller. A digital valve controller is a product for controlling different types of industrial process valves. The two main customer segments for these valve controllers are pulp and paper industry and process and energy industry. According to the company strategy the next generation of valve controllers should be cost competitive, gain technology leadership and be a service platform that includes upgrading, service, software and diagnostics.

2.1 Objectives

The main objective in this case study was to find conceptual modular product designs that would make possible to reduce future development costs for the coming new product generations and to decrease the costs of purchased parts, manufactured parts, assembly and testing to improve competitiveness and to achieve a long term growth.

Some other objectives were to find conceptual designs that would decrease the total product costs in the next product generation, improve the product performance in needed areas and support needed communication standards.

Based on the objectives above, a modularisation strategy for the digital valve controller was defined. The modularisation strategy based on so called module drivers [Stake 1999] specifies which types of modules would be important to achieve. In this case these drivers were *common unit*, *technical specification*, *technical evolution*, *carry-over*, *upgrading*, *separate testing* and *supplier available*. To reduce costs of some components by increasing their volumes, the aim was to find as many common unit modules as possible in the valve controller product family. To limit the effects of variation (due to different technical specification from customers) to some modules, the idea was to create some technical specification modules. Some parts of the product can also be affected by external technical evolution during product life cycle and those effects must also be limited to some modules. To reduce future development costs some technical solutions that will not be changed in the future product generation can be grouped together to form carry-over modules. To make upgrading of the product cost effective some technical solutions or options can be grouped together to form upgrading modules. Some components also can be or need to be tested separately before the final testing of the product and some of those components can be grouped to form separate testing modules. Components that can be (further developed), manufactured, assembled and tested by a supplier can form supplier available modules.

2.2 Outline of the product family

The main tasks done in this case study were market segmentation, clarification of customer requirements in each segment, functional analysis of the product, selection of technical solutions, generation of module concept and creation of product strategy and development plan.

Before starting to specify a modular product platform the needed product variants must be identified. The market segmentation, see an example in case 3, puts the focus on interesting market segments and helps to outline the product family.

In the first step of the method must be ensured that appropriate design requirements are derived from the real customer/market needs. Therefore, in every case, the customer needs must be identified and the most important of them must be clarified. The most important customer needs are then used as an input when developing the modular product platform.

Because in this case the markets were clearly divided in two main segments which had different requirements to product the segmentation tree was created and the customer requirements were clarified separately in both segments.

Before starting to work with the actual product modularity, also the functional structure and the technical solutions of the product family must be clarified. In this case it was done with the help of functional trees to keep the functions of the product as independent as possible. The functional trees are hierarchical trees where the functional structure of a product is first clarified on very abstract level after which the possible technical solutions of the product are added to the tree. Because in this case in the functional trees several technical solutions were found to some functions of the product, some technical solutions were also selected with the help of Pugh selection matrices. In a Pugh selection matrix, the evaluation is made by comparing each alternative with reference that can be for example an existing technical solution.

2.3 Modular product platform

The technical solutions selected were analysed regarding their reasons for forming modules (module drivers) and the module concept of the valve controller product family was generated with the help of Module Indication Matrix (MIM). After filling the MIM, also the module driver profile was generated and the possibilities to group some technical solutions in the same module were studied. The module driver profile of the valve controller family shows clearly how the different reasons for modularisation were weighted in the Module Indication Matrix, see Figure 5.

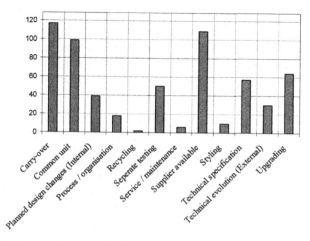

Figure 5: Module driver profile of the valve controller family [Österholm 2000]

As the driver profile brings out, the module drivers carry-over, common unit and supplier available were highly weighted and also the other important drivers were medium weighted in the matrix. That indicates that the module driver profile quite well fulfils the defined modularisation strategy. Based on the generated module concept the product strategy and the development plan of the product family were also created in this case study.

3   Case 2 – Electrical Faucet

In the case 2 the case product was an existing electrical faucet product family that consists of approximately 70 different products. Besides of electrical faucets the electrical faucet product family consists also of electrical urinal flushers and electrical showers. The common feature between all products was the touchless identification of need of water that is done with the photocell.

The products are today mainly used to control water flow in industrial (e.g. food industry and hospitals) and public (e.g. restaurants) applications but the home-use applications are seen to become more important in the future. Also communication with the building automation systems is seen to become more important in the future.

3.1 Objectives

In order to improve competitiveness of electrical products and to make home-use more common there was a need to reduce the price of the products. One way of achieving this price reduction was by increasing the volume of some components that are common between the different products and market segments. For the present generation of products, there has also been problems caused by applications added to the products that had not been considered in the definition phase. These added applications have changed the requirements and thereby caused some different low volume solutions.

The main objective in this case study was to create a concept for a "true" product family that is built from a well defined set of modules. The focus was to have as many common modules as possible and the reason for this was to increase the volume and thereby reduce both direct and indirect costs.

The other objectives were to be able to upgrade/improve modules over the complete range of the product family and to change design without changing the common modules or carry-over modules.

Based on the objectives, the main reasons for modularisation were found. Those module drivers were *common unit, styling, carry-over* and *technical specification*. To reduce costs of some components by increasing the volumes, the aim was to find common units in the product family, and to reduce development costs in the future by using as many carry-over modules as possible. To limit the effects of variation to some modules and not to change common unit modules or carry-over modules there was also a need to find styling modules and technical specification modules.

3.2 Outline of the product family

The main tasks done in this case study were market segmentation, clarification of differences between different segments, clarification of customer requirements to different product groups (faucets, urinal flushers and showers), functional analysis, selection of technical solutions, generation of an ideal module concept, evaluation of the ideal module concept and creation of a development plan.

Because in this case the markets were divided on the other hand in three segments (industrial, public and home-use) but also in three product groups (faucets, urinal flushers and showers), the main differences between different segments and different product groups were clarified before clarification of customer requirements. After that the most important customer requirements and stakeholders to each segment and product group were clarified separately. In order to know the importance order of the requirements the project group made the ranking with the help of so called Pairwise matrix, see Figure 6. In a Pairwise matrix the comparison of requirements is done simply by comparing each requirement one by one to other requirements. In the matrix one requirement is either more important (+) or less important (-) than the other what gives the final importance order when the whole matrix is completed.

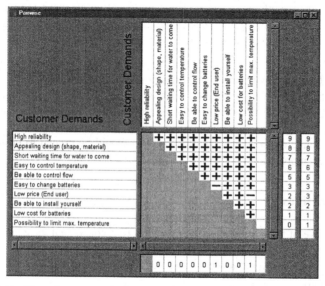

Figure 6: Ranking of customer requirements with the help of the Pairwise *matrix [Öster-holm 2000]*

After clarification of customer requirements, the functional structure of the products was also in this case analysed with the help of functional trees and all different technical solutions were listed to be able to see which technical solutions fulfil the same function in different products.

3.3 Modular product platform

Because the case product was an existing product family that consists of approx. 70 different products and three different product groups, it was not possible to find many common unit modules that are common in all products of that product family. To be able to continue, some of the technical solutions were selected and an ideal module concept was generated with the help of Module Indication Matrix.

The ideal module concept was also evaluated in this case study and some new ideas, like how the product family can be changed and unified in the future, were found. Based on these new ideas the development plan for the electrical faucet product family was created.

4   Case 3 – Air Valve

The case 3 deals with an air valve. The function of an air valve is to control the airflow in different air ventilation systems. Industrial factories, offices and public buildings are potential application areas for using air ventilation systems.

This case started with an idea of having a textile material tube construction of the air valve. Use of a textile material would possibly facilitate noise and cost reductions. Since the present air valve design (see figure 2) is used in a range of products so this would be also the case for the textile valve design and a modularised design should support the scalability and reuse of components/modules. If a complete new range of air valves were designed it would be crucial to ensure a systematic design procedure.

4.1 Objectives

The objective of this case was to evaluate different conceptual ideas, make a proposal for a modular product concept and outline a verification plan for recommended concept/concepts to be carried out in the subsequent phases of product development.

Based on the objectives above, a modularisation strategy for the tube air valve was defined. The modularisation strategy based on so called *module drivers* [Stake 1999] specifies which types of modules would be important to achieve. In this case these drivers were *common unit, technical specification, separate testing* and *carry-over*. To reduce both the costs, number of suppliers and hopefully also to increase the volume of some components the aim was to define some modules that shall be common units in the product family. In this case we tried to collect components that are subject to the technical specification into some modules and thereby decrease the "adjustments" and the time needed from order to delivery. The idea is to limit the variations (due to different technical specifications from customers) to some modules. Separate testing means that we tried to collect some components so that we could do some of the testing before final assembly and thereby cut some time in the assembly. To decrease the time needed from order to delivery the project team tried to increase the number of carry-over modules and thereby reduce the time needed for product adjustments.

4.2 Conceptual ideas

Only one conceptual idea existed in the initial phase of the project. If a pre-study is based only on one or two different concepts, there is a risk that good ideas do not come into consideration. Therefore alternative solutions were brainstormed and documented. The project team came up with eight different sketches. One of those sketched prototypes is presented in Figure 7. In this sketch the inner diameter of the air valve is changed both by rotating and moving the textile material inside the tube [Uuttu 2000].

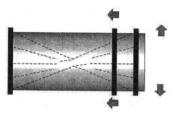

Figure 7: A sketched prototype. In this sketch the diameter of outgoing air is changed both by rotating and moving the textile material inside the tube.

4.3 Outline of the product family

Before starting to specify a modular product platform the product variants must be identified. Market segmentation, see Figure 8 [Uuttu 2000], puts the focus on interesting segments and helps to outline the air valve family. With a market segmentation tree we can find answers to such questions as how big is the market and which market segments would the products be desirable to cover?

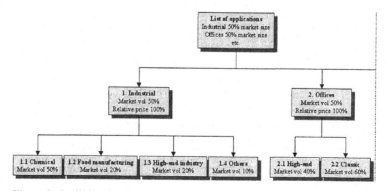

Figure 8: Outlining the air valve family – an example of market segmentation

Referring to MFD [Erixon 1998] different technical solutions must be identified during the modularisation work. This was done by the help of a functional tree.

4.4 Modular product platform

The technical solutions selected were analysed regarding their reasons for forming modules. The criteria used are the *module drivers* described before. This was done with help of a matrix type working method called MIM (Module Indication Matrix). Each technical solution is assessed against the module drivers. In Figure 9 the MIM of the tube air valve is presented. There every technical solution is weighted on a scale where nine points correspond to a strong driver, three points a medium driver, and one point a weak driver, according to the importance of its respective reason for being a module. In Figure 9 the module drivers were modified to be better suited to this specific product and case. Based on this matrix the modular product platform was specified, see Figure 9. In Figure 9, at right hand side, each module has an unique reason for being a separate module. In the base module some of the technical solutions that had strong features of both *Planned design changes (internal)* and *Technical specification* drivers were grouped together to a module. Working like this the aim was that future design changes and product variation affect only few modules. The sealing module has an analogous modularisation strategy when comparing to the Base module. In a similar way some of the technical solutions which had a *carry-over* and *common unit* module pattern were grouped together to form a Screw and clinch module. This means that we had separated parts from the product structure to one module, parts which are common in the product family and most likely will not be exposed to any design changes during the life of the product platform.

Technical Solutions	Carry Over 5-10 yrs	Planned Design Changes (internal)	Variants	Common Unit	Process / Organisation	Strategic Supplier Available	Service / Maintenance	Recycling		Modules
	54	63	52	48	2	9	-9			
Base		9	9		1			1	19	Base module
Guide Pin		9		9				1	18	
A round circle	9		9					1	18	
Screw connection	9			9				1	18	Screw and clinch
Clinchs connection	9			9				1	18	
Pressure holes	9			9				1	18	
Gasket		9	9					1	18	Sealing module
Seal		9	9					1	18	
Connection parts	9		9		1			1	19	Important variants
Adjusting lever	9		1	3				1	13	Two variants ?
The textile material		9	3			9	-9	3	12	Separate
Spring		9	3					1	12	Separate
Piano Wire		9		9				1	18	Separate

Module Drivers

Figure 9: Module Indication Matrix (MIM) of the air valve

5 Case 4 – High Speed Assembly Cell

The case study 4 concentrates on HiSAC® (High Speed Assembly Cell) products. HiSAC® products are solutions for the end line production automation mainly for the electronics manufacturing industry. They are designed to perform different types of electronic manufacturing related automation operations, such as odd-form component assembly, final assembly and PCB routing.

5.1 Objectives

The main target for this project was to define a modular product family that would help us to decrease the time needed from order to delivery. One important target was also to establish the same view of the module concept (modular product platform) between the development, sales and production competencies.

Based on the objectives above, a modularisation strategy was defined for the Hi-SAC® products. In this case the selected module drivers were *common unit, technical specification, separate testing* and *carry-over*. To reduce both the costs, number of suppliers and hopefully also to increase the volume of some components the aim was to define some modules that shall be common units in the product family. In this case we tried to collect components subject to different technical specification into modules and thereby decrease the "adjustments" and the time needed from order to delivery. The idea is to limit the variations (due to different technical specification from customers) to some modules. Separate testing means that we tried to collect some components so that we could do some of the testing before final assembly and thereby cut away some time in the assembly. To decrease the time needed from order to delivery the project team tried to increase the number of carry-over modules and thereby reduce the time needed for product adjustments.

5.2 Outline of the product family

Before starting to specify a modular product platform the product variants must be identified. Market segmentation, see an example in case 3, puts the focus on interesting segments and helps to outline the HiSAC® family.

Referring to MFD [Erixon 1998] different technical solutions must be identified during the modularisation work. This was done by the help of a Functional tree.

5.3 Generate concepts

The technical solutions selected are analysed regarding their reasons for forming modules. This was done with help of the matrix type working method MIM (see an example in case 3). The criteria used are the module drivers described before. In Figure 10 [Uuttu 2000], the old and new module driver profiles are compared. As one can see, in the old HiSAC® structure there were many systems/modules/ areas which are strongly affected by the technical specification driver and therefore this driver profile has the largest value. On the other hand there were very few systems/modules/areas which are common or have common solutions in the HiSAC® product family. In the new driver profile the commonality opportunities are estimated. It shows that there is a great potential in finding common solutions in the entire HiSAC® product family and at the same time the effect of different technical specifications can be minimised into some modules only.

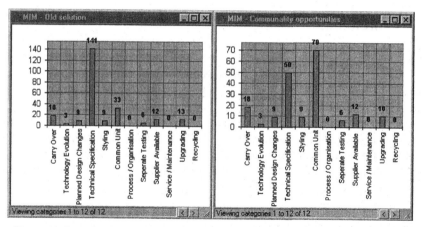

Figure 10: Estimating communality opportunities of the High Speed Assembly Cell

With the help of the Module Indication Matrix [Erixon 1998] an analysis of the module drivers for the different systems/modules/areas was conducted (see Figure 11). This analysis gives an indication of which systems/modules/areas have strong or very strong reasons for forming a separate module area. As a result of this analysis, the electrical cabinet, the main cables etc., the chassis and the covers seem to present strongest reasons for forming a separate module area.

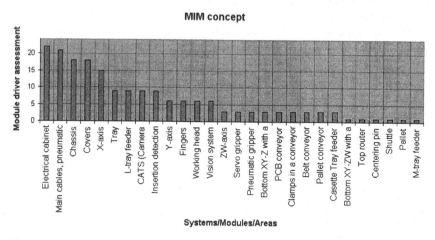

Figure 11: Analysis of which systems/modules/areas present strong reasons for forming a separate module area

6 Conclusions

Based on these four case studies [Uuttu 2000] and [Österholm 2000] the applied Modular Function Deployment (MFD) method seems to be a good tool for structuring the working process in the early phases of product development. The systematic way of working in the MFD method decreases the market related risks and drives the project team to think different customer requirements and market segments, needed product variants and possible future changes that can be either planned (internal) or unplanned (external).The MFD method also increases cooperation between different functions of the company and helps the project team to use cross-functional know-how.

The output of the product development process based on the MFD method is a well defined product platform from which all the product variants required can be manufactured. In order to succeed with the modular product also in the future, all those in the organisation working with product structure must understand the new way of thinking and the structure of organisation must support modularisation. Only in this way the defined interfaces between different modules can be retained the same for a period of time so that the benefits of modularisation can be achieved. This means also that the modularisation project has to have full support from the management team.

The experience from these four case studies also shows that if some new technology and new principles are to be introduced in a product they need to be verified before starting to work with the modular concept generation. Development of a modularised product family is much simpler than development of the complete product range.

Based on the experiences in the four discussed case studies, the use of a systematic working method supporting efficiently modularisation is highly recommended as part of a concurrent development process of new products. In order to succeed with modularisation the project team must think in terms of product families instead of single products. Only then the risk of expanded product complexity can be minimised when new product variants are introduced.

References

[Erixon 1998] Erixon, G., 1998: Modular Function Deployment – A Method for Product Modularisation. Dr. Thesis, The Royal Institute of Technology, KTH Högskoletryckeriet, Stockholm, Sweden. [178 p.]

[Stake 1999] Stake, R., 1999: A hierarchical classification of the reasons for dividing product into modules – A theoretical analysis of module drivers. Lic. Thesis, The Royal Institute of Technology, KTH Högskoletryckeriet, Stockholm, Sweden. [65 p.]

[Uuttu 2000] Uuttu, O., 2000: Systematic Concurrent Design of a Modular Product Family and Assembly System. M.Sc. Thesis, Tampere University of Technology, Institute of Production Engineering, Tampere, Finland. [93 p.]

[Yxkull 1999] von Yxkull, A., 1999: Modularization & Product platforms. Light Assembly Industry -technology programme, Espoo, 26.8.1999, Finland. 36, 39.

[Österholm 2000] Österholm, J., 2000: Modular Product Structure and Supporting Systematic Design Methods (in Finnish). M.Sc. Thesis, Tampere University of Technology, Institute of Production Engineering, Tampere, Finland. [98 p.]

[Österholm 2001] Österholm, J.,Uuttu, O., Tuokko, R., 2001: Systematic Design of a Modular Product Platform – Applying the Modular Function Deployment Method in Light Assembly Industry. Report 53. Tampere University of Technology, Institute of Production Engineering, Tampere, Finland. [41 p.]

Part III

A Value Chain

Mass Customization and Beyond – Evolution of Customer Centricity in Financial Services

Winter, R.

Summary: For about a decade, mass customization is discussed as a generic methodology to individualize products and services while preserving cost-efficient production processes. Particularly for information-oriented industries like financial services, the advent of electronic business may create new opportunities to individualize products and services. In this paper, a product-oriented and a process-oriented approach to individualization of financial services are presented: On the one hand, by adopting configuration methods from mechanical engineering, financial services can be generated according to individual customer needs. On the other hand, by separating a production-oriented 'factory' layer and a customer process-oriented 'integration' layer in business networks, individual consumer processes can be supported holistically.

1 Introduction

Mass customization means "that the same large number of customers can be reached as in mass markets of the industrial economy, and simultaneously they can be treated individually as in the customized markets of pre-industrial economies" [4, 169]. It is often attributed to the post-industrial age or information age that consumers set the rules and that, as a consequence, even a greater extent of individualization must be realized in an efficient form [9]. "Being truly customer focused is not possible if the organization is not, first, information intensive" [3, 9].

In this paper, different concepts for customer centricity and service individualization in financial services are presented. Based on a short analysis of emerging business models and business architectures of the information age as well as of conceptual foundations of customer centricity (Section 2), a product-oriented individualization approach (Section 3) and a process-oriented individualization approach (Section 4) are presented. The paper closes with conclusions and a short outlook in Section 5.

2 Customer Centricity in the Information Age

The most important business potential of IT innovations is that an advanced technical networking infrastructure together with an organizational business networking infrastructure allow to broaden the transformation scope from an isolated company (or a company-centric view including direct customers and direct suppliers) towards entire value networks.

Parallel to the networking sophistication in the business arena, internet literacy and internet access are available to a growing portion of end consumers. By communicating directly with producers / service providers via electronic channels, end consumers can be integrated into value networks more closely, and product / service design can be influenced more directly. Particularly for products bound to electronic media (e.g. information, music, movies, games, software) or financial products (e.g. loans, investments, insurance), the value network can be separated into a production-oriented 'factory' and a consumer process-oriented 'solutions' portion [16]. Since missing physical restrictions allow for a flexible combination of such products and services, business models can be tailored to specific needs of customer segments instead of having to follow the structure imposed by production processes.

Hence, two arenas contribute to the transformation from traditional industry structures into value networks: [16]

- Business-to-business arena ('business networking'): definition of inter-company business processes, creation of shared services, flexible sourcing, emergence of electronic marketplaces

- Business-to-consumer arena ('electronic commerce'): direct access to (and direct response to) products and services by end consumers via electronic channels, separation of production oriented business models and customer process oriented combination of products and services

2.1 Architecture of Business Networks

Value networks comprise nodes that represent certain roles which do not exist in traditional business architectures or at least are not elaborated to that extent: [16]

- Customer processes become the focus of the value network design. *Service integrators* combine product and service components produced by different service providers to create solutions that are tailored to a specific, holistic customer process or a specific life event. In most cases, information components become a more important solution component than in traditional products.

- While most products and services are used by several service integrators, some products or services may be produced for only one or few service integrators (or service providers in later production stages). By providing such services, *exclusive service providers* contribute to an unique selling proposition.

- While exclusive service providers are focussing on exclusivity, *shared service providers* produce large amounts of standardized products or services for use by several other service providers or by service aggregators. Shared service providers focus on mass customization and economies of scale.

- While it may be useful for exclusive service providers to maintain direct, exclusive networking links to service integrators, shared service providers need to use a common networking infrastructure for doing business with a large number of other service providers and service aggregators. This common infrastructure, the *business bus*, allows (shared) service providers and service integrators to exchange electronic products, electronic services, and information. To allow for efficient business networking, a large number of standards is defined not only on the lower, communication oriented layers of the business bus (e.g. EDIFACT, XML based standards), but particularly on the higher, business oriented layers (e.g. contracting standards, service level agreements).

- The business bus is complemented by several basic (electronic) services that are industry independent or process independent (e.g. certificates, payments, creditworthiness information). These *public services* are closely linked to the business bus.

In Figure 1, a general architecture of value networks is illustrated.

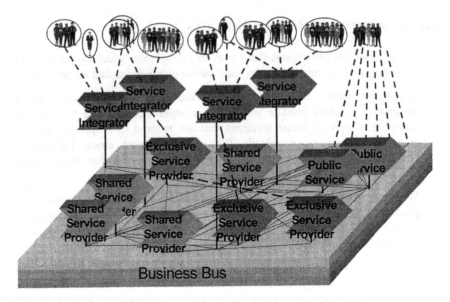

Figure 1: Business Architecture of the Information Age [2]

Most end consumers have access to products and services using service integrators. However, if end consumers wish to aggregate products / services by their own, they can access the business bus, too.

All types of service providers optimize their business models with regard to their production processes. In contrast, service integrators optimize their business models with regard to customer processes or customer life events, respectively. Business networking and the utilization of the business bus allow to separate production-oriented and customer process-oriented portions of the value network, thereby allowing different business models to exploit their respective core competencies.

2.2 Conceptual Foundations of Customer Centricity

Customer centricity approaches often fail due to their focus on technical solutions (e.g. customer relationship management systems) although a holistic approach that covers the conceptual foundations would be needed [14]. But what are the conceptual foundations of customer centricity? A multi-layered specification approach that has proved useful in a large number of research and industry projects [16] is illustrated by Figure 2.

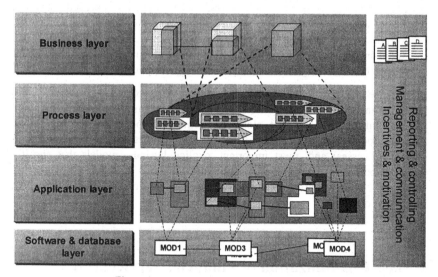

Figure 2: Modeling layers for Value Networks

While products and services, distribution channels, markets, core competencies as well as other business-related properties are specified on a business layer, respective business processes are specified on a process layer, supporting applications are specified on an application layer, and respective software modules and database components are specified on a software & database layer. While several abstraction levels are usually used on every layer to represent the respective specifications in different detail, different layers are connected by transformation rules. While each layer is used to represent respective "hard facts" formally (i.e. based on at least semi-formal meta models and modeling methods), soft facts do usually not belong to one distinctive layer nor can be represented formally. As a consequence, a separate, document based layer is used to represent this aspect.

Systems implementations (e.g. CRM applications) are specified using some systems development paradigm on the software & database layer. These specifications, however, have to be guided by a respective application architecture which, in turn, depends on the specification of customer processes which, at the end, are supporting a certain business model. For production oriented business models as well as for customer process oriented business models, products or services, supported distribution channels, targeted markets, targeted customer segments, critical success factors, etc. have to be specified first. In a second step, appropriate business processes can be specified that create the specified outputs in an efficient and controllable way. On this foundation, an appropriate application architecture can be specified in a third step which defines effective systems integration and decoupling. Finally, in a fourth step software modules and database components can be configured to support the target application architecture.

As a consequence, the conceptual foundation of successful customer-centric solutions consists of an appropriate application architecture (e.g. integration along customer information and not along platforms or product lines), appropriate business processes (e.g. relationship management processes and not transaction oriented processes) and, most important, explicit business models that specifiy which customer processes for which customer segments (and / or for which life events and / or in which life phases) are supported by which services over which distribution channels.

3 Mass Customization of Financial Services – A Product-oriented Approach

If complex products are manufactured, a model based configuration of product variants has many benefits with regard to product data management, production flexibility, and systematic resource management. Hence, product models and product configurators have been incorporated into standardized production management software packages (e.g. SAP R/2 or R/3) and are also a common feature in individual production and / or sales support solutions of most investment goods manufacturers. An overview of configuration concepts can be found in [7]. Although complex products are also created by banks and insurance companies, a model based configuration of product variants can only be recently found in standardized software packages (e.g. FJA's or Alldata's solutions [1]). In individual software solutions and even some standardized software packages, product data are usually 'hard-coded', i.e. product structures and feature dependencies are hidden in tariff systems, 'individual' surcharges or reductions, and sales bundles [12]. An important reason for the lack of product orientation of most current software solutions is that product handling is distributed over numerous business unit related application systems (e.g. life insurance, health insurance, car / home insurance) and / or over numerous business process related application systems (e.g. policy management, claim management, premium management). During the last decade, model based configuration of products has been adopted by many financial service companies to separate business unit support and / or business process support from product data (and tariff) management, thereby enabling companies to implement product (and tariff) innovations more consistently, more flexibly and, what maybe most important, in much less time [18].

The conceptual basis for configuration of individualized products in financial service industries is presented in subsection 3.1. Subsection 3.2 illustrates the configuration process using an example from the insurance industry. To create product variants (and standard products) from general product models, a certain set of production rules is applied. If these rules are inverted, it should be possible to link product variants (or standard products) back to the underlying, generalized product models. Such an 'inversion' of the configuration process allows for linking product variant related information (e.g. profitability) to the underlying product model, thereby guiding the product innovation process to focus on profitable products and to avoid unprofitable variants. The 'inverse' application of configuration rules is described in subsection 3.3. The contents of this section have been initially developed in [12] and been elaborated in [13].

3.1 Configuration of Product Variants

Product configurators are usually utilized to derive product variants as a combination of product components and / or by checking the feasibility of given combinations against a generalized product model [7]. A more powerful concept that underlies state-of-the-art standardized software packages for mechanical engineering companies is the concept of open variants [20, 124-128]: Product variants are specified by certain values for a set of attributes that depends on the respective product type. This results in a three-level hierarchy: Product types are assigned to attributes, and attributes are assigned to attribute values. Attributes may be compulsory or optional. Moreover, attributes may be single-valued (i.e. attribute values are exclusive) or multi-valued. Furthermore, values of different attributes may be linked by dependencies (inclusion or exclusion).

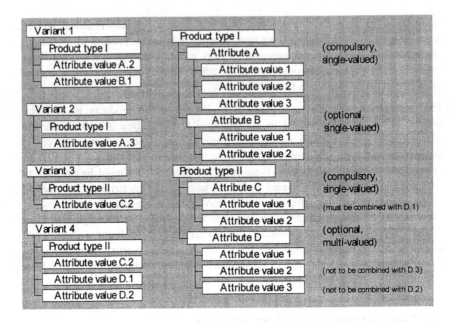

Figure 3: Derivation of Product Variants (adapted from [13])

Based on an open variant model (right side), the derivation of four product variants (left side) is illustrated by Figure 3. The simplified product model comprises two product types I and II which are specified by two attributes sets A B and C D, respectively. While A and C are compulsory, B and D are optional. A, B, and C are single-valued attributes, while D may have more than one attribute value. However, if C.1 is chosen, then D.1 must also be chosen. Furthermore, D.2 may not be combined with D.3. This simple product model implies a total of 13 feasible product variants.

The simplified product model for open variants may be elaborated by various types of additional constraints, creation of tables for maintaining feasible combinations of attribute values, and integration of individual calculation routines.

3.2 Application Example: Insurance Products

If configuration concepts from mechanical engineering are to be applied to financial services, two specialties of services must be taken into consideration:

1   Not all product variants may be generated that are technically feasible. For services, it is essential to integrate customer properties into the configuration process [12, 33-34].

2   While maximum flexibility has been the foundation of product models in mechanical engineering, service product variants primarily reflect regulations, pricing rules, customer properties, or risk properties. Therefore, more constraints have to be represented in general.

The primary advantage of an open variant product model for financial services is a simple representation of configuration rules.

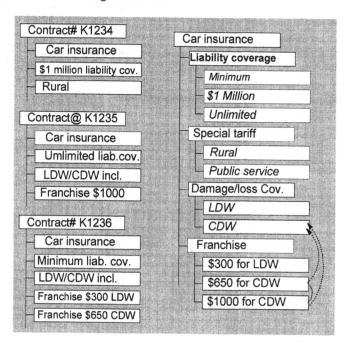

Figure 4: Insurance Contracts as Open Variants

Figure 4 illustrates the application of the open variant product model to financial services: Based on a general model of car insurance contracts (right side), the derivation of three concrete contracts (left side) is shown. The notation is condensed from Figure 3 by denoting compulsory attributes in bold font (vs. optional attributes in normal font), values of single-valued attributes in italic font (vs. values of multi-valued attributes in normal font), exclusion rules as undirected arcs, and inclusion dependencies as dotted, directed arcs. A detailed analysis of the applicability of open variant product models to different insurance and banking service types can be found in [12][13].

3.3 Inverse Configuration

Usually, product models are utilized to generate variants based on a general product structure and a set of constraints. But it is also possible to utilize a product model in an inverse way: Since the generation rules for variants based on the general model are known, information related to concrete variants can be transformed into information related to product types, attributes, and attribute values. E.g., using the product model illustrated by Figure 4, profitability information for insurance contracts could be transformed into profitability information linked to the various liability coverage, tariff, damage / loss coverage, and / or franchise alternatives. This information is particularly valuable to support pricing decisions or decisions regarding product structure.

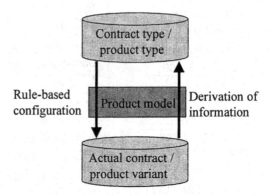

Figure 5: ‚Forward' and ‚Inverse' Application of Product Models (adapted from [13])

Figure 5 illustrates the two different utilization directions of an open variant product model: While the usual configuration type utilization creates product variants from a general model of product types, its inverse utilization derives type level information from variant information.

4 Intentions Value Networks in Financial Services – A Consumer Process-oriented Approach

Consumers are looking for ways to manage basic needs and achieve basic objectives in a simplified way [6]. As an example, 'traveling' today often means that travel products (e.g. flight, accommodation), banking products (e.g. foreign currency), health products (e.g. vaccination), insurance products (e.g. baggage insurance), and information products (e.g. travel guide) have to be purchased from different 'factories'. This situation is illustrated by Figure 6. More complex objectives like planning for retirement, moving to a new community, or starting a new career require even more products and services to be integrated. Such complex life objectives have been designated as consumer 'intentions' [6].

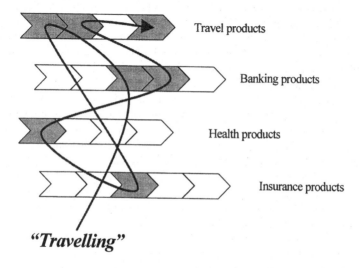

Figure 6: Intention 'Traveling' vs. Traditional Travel Related Products and Services [2]

Since deregulation and globalization induce not only more choice, but also more complexity in consumer's buying decisions, isolated products and services lose their traditional importance. When buying is more and more driven to create a holistic solution to a problem or to entirely satisfy an intention, companies which want to succeed have to simplify consumer's processes, i.e. by integrating products and services for focused customer segments.

The conceptual basis for consumer process-oriented design of business models is presented in subsections 4.1 and 4.2 from the consumer's and the provider's perspective, respectively. The contents of these subsections have been initially developed in [2]. Subsection 4.3 uses an example from retail banking to illustrate consumer process-orientation.

4.1 Modeling Consumer Processes – The Consumer's Perspective

The consumer process from the consumer's point of view is based on a multi product mix which defines itself according to the specific life event or intention that triggers the buying process. At the beginning of the buying process, the consumer does not always know the individual products which constitute the final product offered. Furthermore, this product does normally not have a company-related brand so that consumer loyalty is difficult to establish and the consumer lock-in process must be redesigned [19, 103-171]. The problems we face for branding an intention-triggered product can be compared to the difficulties which occur when establishing a brand for a 'traditional' service product. A physical product, e.g. a specific car, 'transports' a brand and thus is recognized by the consumer due to its potential characteristics. A consulting company establishes a brand which is closely tied to its name, rather than its products. In this case the branding process is much more difficult, because the product cannot be directly recognized by the consumer. The product is defined by a good measure of imagination and suggestive power of the advertising campaign [10, 44-45]. The same is true for an intention-triggered product: From the consumer's point of view, the lock-in is achieved by a product mix, which satisfies the intention in the best possible way, i.e. the most important characteristics of the product are its quality and degree of individuality. First experiences of Web based aggregators show that 'true choice' and 'objective advice' become more important constituents of consumer loyalty than factors that can be attributed to isolated products [11].

The consumer process, which results from the above discussed points, can be defined as follows:

- Identification of an intention

- Search for an aggregator which offers not only appropriate solutions, but also 'true choice' and 'objective advice'

- Selection of appropriate solutions including desired distribution and access channels

- Payment and 'consumption'

4.2 Modeling Consumer Processes – The Provider's Perspective

Following an outside-in strategy, the consumer process from the provider's point of view must be defined according to the process described in the preceding subsection. This leads to the following provider's process steps:

- Identification and selection of communities / consumer segments. While traditional approaches are based on income, geographical, educational, or similar segmentation, alternative approaches include an analysis of individual attitudes [5].

- Anticipation of intentions. This analysis can be done using an event-oriented approach ('Which life events are triggering buying decisions?'), a life stage-oriented approach ('In which life stages, which buying decisions are made?', 'Which life stages precede / succeed which life stages?'), or a combination of both.

- Selection of 'best of breed' solution components. In order to provide 'true choice' and 'objective advice', the network should be large, and selection decisions should be made based on quality assessment rather than be based solely on profitability.

- 'Assembly' of solutions which contribute to the satisfaction of one or more intentions.

- Support of all appropriate distribution and access channels. While maximum channel usage flexibility should be maintained in general, assignment decisions can be based on a content model [8].

- Obtaining and keeping a competitive position in the network. For maintaining a favorable position in a value-added network, the concept of networkability [15] has been proposed.

If a company is a component of an intention value network, it contributes to the final solutions by providing components rather than being the last element of the supply chain. Thus, several changes and effects result:

The consumer is no longer a 'holistic' marketing object, but is segmented according to his or her intentions. The company must nevertheless know the intentions and the possible solutions in order to be able to produce the 'right' product or service component. Customer relationship management in this context means that the service integrator as direct interface to the consumer has to individually care for him or her. As long as the product was the link to a 'faceless' mass of median consumers, it could be manipulated according to more or less fuzzy results of empiric research. Today, the connection is directly established with the consumer and the care must be 'applied' according to his wishes. That means that the product has to be individually customized for a specific consumer. Otherwise, since the switching costs are very low, the consumer can easily move on to another integrator.

Customers from the companies' point of view can be divided into three groups:

1. End consumers which buy and combine products and services directly (without using a service integrator)

2. End consumers which buy solutions, i.e. which use a service integrator

3. Service integrators

Since, in the presence of service integrators, a direct contact to end consumers becomes increasingly difficult for producing companies, differentiation as far as competitors are concerned must be created by other means. This can, for example, be done by banner advertising on the respective Web site, targeting particularly customers in groups one and three. A prerequisite for being able to do this is that the company stays in the network. Thus, on the one hand, the company has to keep a direct contact to the service integrator and create a lock-in position with it. On the other hand, and this must be the main ambition, the entire organization must be orientated towards keeping the position in the new business architecture of the information age, which is developed in the next chapter.

Another important point of the multi product mix is that it is constituted by products of different companies. Therefore we need a reliable standard for enabling the intermediaries [9] (service integrators) to assemble the final solution. Service integrators must be able to integrate product and service components by using a generic method in order to achieve a maximum degree of individuality. Generic products can be integrated by using product catalogues [17] which serve as product pool for creating individual solutions.

4.3 Application Example: Home Buying Process

Among the pioneer intention value networks created in 1999 was ThirdAge.com. ThirdAge.com 'presents the best the Web has to offer for active older adults. ThirdAge.com offers rich advice and helpful information on topics critical to the

ThirdAge community, such as retirement housing and estate management - areas typically ignored or glossed over by traditional media.' (http://www.thirdage.com) By integrating services from Drugstore.com, E*Trade, GE Auto Insurance, IBM, Intel Relocator, Merrill Lynch, Millstone Coffee, etc., and by partnering with Barnesandnoble.com, Ecentives, Global Health and Fitness, Monster.com, Softwatch, Weather.com, Women's Health In-teractive, WorldRes, etc., ThirdAge.com tries to satisfy the intentions 'achieve financial security', 'maintain a healthy lifestyle', 'find appropriate housing', 'expand personal horizons', and 'explore the world'.

In the field of financial services, Credit Suisse, the retail banking unit of Swiss Credit Suisse Group, has launched a portal in 2000 that is intended to support end customer's home buying processes in a holistic, product-independent way. Yourhome.ch integrates geographical and statistical information with services e.g. from real estate brokers, architects, interior designers, tax consultants, legal consultants, real estate consultants, moving companies, insurance companies, and of course mortgage banks. In addition to general information and product-related information, among others a utility to locate an address on a map and a utility to interior design is provided. Of particular interest is the fact that, although the portal is provided by a bank, it is planned to integrate products of competitors.

5 Conclusions and Outlook

We have presented two different individualization approaches of financial services. While the configuration approach allows to generate highly individualized product variants based on a product model that comprises a large number of attributes and attribute values, the intention value network approach caters to specific consumer intentions by integrating a large set a products and services with relevant, objective information.

- The configuration approach represents an inside-out view to individualization: Corporate product development tries to reflect consumer needs by offering customizable services within the scope of a 'factory' oriented business model.

- In contrast, the intention value network approach represents an outside-in view to individualization: Based on an in-depth analysis of consumer processes, corporations with complementary core competencies form a business network to support consumer processes holistically.

Both approaches seem to be useful to support an open finance setting: On the one hand, service integrators need flexible 'product factories' producing customizable products efficiently. On the other hand, product factories need service integrators that provide effective customer knowledge and effective customer care.

Current developments in the retail banking sector support that hypothesis: Universal banks disintegrate into product factories that produce shared services (e.g. transaction banks, online brokers, card services), exclusive service providers (e.g. risk management, investment consultants), and service integrators (e.g. home buying specialists, retirement specialists). Simultaneously, banks begin to collaborate with traditional customer knowledge / customer care specialists like insurance brokers, food retailers, airlines, or consumer portals.

If customers increasingly use integrators for process support in specific life stages and / or life events, however, the question arises whether branding and loyalty-based customer management are sustainable approaches. Long-term and holistic consumer relationships seem to be contradictory with a life event / life stage-oriented, core competency-based approach. Moreover, it is in question whether the integrated data needed for customer knowledge management will be available to an integrator, a network of service providers, or only the consumer.

References

[1] ALLDATA SDV GmbH (1998): CAD/IP Computer Aided Design for Insurance Products.

[2] Baumöl, U., Winter, R. (2001): Intentions Value Networks – A Business Model of the Information Age. In: Miranda P., Sharp, B., Pakstas, A., Filipe, J. (Eds.): ICEIS 2001 – Proc. of the Third International Conference on Enterprise Information Systems, vol. 2. Setubal: ICEIS Press 2001, 1075-1080.

[3] Blattberg, R., Glaser, R. (1994): Marketing in the Information Revolution. In: Blattberg, R. et al. (Eds.): The Marketing Information Revolution, Boston, 9-29.

[4] Davis, S. (1987): Future Perfect, Reading.

[5] Fridgen, M. et al. (2000): Kundenmodell für eCRM – Repräsentation individueller Einstellungen. Research Report WI-79, Institute of Business Administration, University of Augsburg.

[6] Friedman, J.P., Langlinais, T. (1999): Best intentions: A business model for the eEconomy, Outlook, 11, 1, 34-41.

[7] Kleine Büning, H., Stein, B. (1993): Entwicklung von Konfigurierungssystemen. In: Kurbel, K. (Ed.): Wirtschaftsinformatik '93, Physica, Heidelberg, 287-302.

[8] Kundisch, D. et al. (2000): Ein Contentmodell für den Multichannel Vertrieb von Finanzdienstleistungen, Research Report, Institute of Business Administration, University of Augsburg.

[9] Hagel, J., Singer, M. (1999): Net Worth – Shaping Markets When Customers Make the Rules, Harvard Business School Press.

[10] Hamilton, Joan O'C. (1999): Selling the I-way on the Highway, Business Week E.Biz, November 1, 1999, EB 44-EB 45.

[11] Langlinais, T., deLeon, A. (1999): Winning the e-economy with "Intention Value Networks", EIU Strategic Finance, The Economist Intelligence Unit Ltd., 50-68.

[12] Leist, S., Winter, R. (1994): Konfiguration von Versicherungsleistungen, Wirtschaftsinformatik, 36, 1, 45-56.

[13] Leist, S., Winter, R. (1998): Nutzung generischer Produktmodelle im Finanzdienstleistungsbereich am Beispiel des Ergebniscontrolling, Wirtschaftsinformatik, 40, 4, 281-289.

[14] Light, B. (2001): A Review of the Issues Associated with Customer Relationship Management Systems. In: Proc. of the 9th European Conference on Information Systems, http://ECIS2001.fov.uni-mb.si.

[15] Österle, H., Fleisch, E., Alt, R. (1999): Business Networking. Shaping Enterprise Relationships on the Internet, Springer, Berlin.

[16] Österle, H., Winter, R.: Business Engineering, Springer, Berlin etc. 2000

[17] Schmid, B., Selz, D., Sing, R. (1997): Electronic product catalogues. EM – Electronic Markets, http://www.electronicmarkets.org/netacademy/publications.nsf/all_pk/717.

[18] Schönsleben, P., Leuzinger, R. (1996): Innovative Gestaltung von Versicherungs-produkten: flexible Industriekonzepte in der Assekuranz, Gabler, Wiesbaden.

[19] Shapiro, C., Varian, H. (1999): Information Rules, Harvard Business School Press.

[20] Zimmermann, G. (1988): Produktionsplanung variantenreicher Erzeugnisse mit EDV, Springer, Berlin.

Modularity in Three Dimensions: A Study of Mass Customization in the Dutch House Building Industry

Wolters, M. J. J., van Heck, E. & Vervest, P. H. M.

Summary: Modularity is often considered as the major enabler of mass-customization. This paper introduces modularity in three business dimensions: products, processes and supply chains. The different aspects and opportunities of three-dimensional modularity are investigated in relation to successful mass-customization strategies. The central proposition of this paper is that a network of organizations will be more effective in pursuing a mass-customization strategy when all three dimensions are concurrently designed in a modular fashion. This proposition is validated within the Dutch house building industry, which is currently trying to mass-customize its products. In particular, we focus on an innovative project, called Dwelling on Demand, which is carried out in the city of Almere. This project is one of the first steps the Dutch housing industry takes towards more customer-influence on a serial basis. It is concluded from this project that a concurrent design in all three dimensions often leads to better mass-customization performances, while it assures a better fit between serving the customers' requirements and the organizational network structure and capabilities.

Keywords: mass-customization, modularity, business networks, building industry

1 Introduction

This paper investigates the different aspects and opportunities of modularity in interorganizational business networks, which are trying to achieve mass-customization effectively. Modularity is by many theorists and practitioners considered as *the* enabler of mass-customization (e.g. Pine 1993). Most attention in this respect is paid to modular design of products. We add two other business dimensions to this perspective: business processes and supply chains, together making up a three-dimensional modularity perspective. It is investigated whether interorganizational business networks, which want to offer mass-customized products to their customers, do in fact modularize in these three dimensions. Do they not only modularize products, but modularize processes and supply chains as well? We try to find theoretical and empirical evidence of the analogy between the three dimensions of modularity.

The central proposition in this paper is that an interorganizational network will be more effective in pursuing a mass-customization strategy when all three dimensions are concurrently designed in a modular fashion. The proposition is validated within the Dutch house building industry, which is currently trying to mass-customize its products. In particular we focus on an experimental project, called Dwelling on Demand[1], which is carried out in a Dutch city called Almere. It is one of the first steps the Dutch housing industry takes towards more customer-influence on a serial basis. It is an experiment to explore the organizational and technical difficulties belonging to such a mass-customization strategy.

The paper is structured as follows. In the following section we elaborate on the use of modularity for the design of products, processes and supply chains. Subsequently, we introduce the Dwelling on Demand case study in the Dutch housing industry and present the findings regarding our central proposition.

2   Modularity in Three Dimensions

2.1   Modularity for Mass-Customization

Sanchez (1999) summarizes four properties and strategic uses of modular architectures:

- Modularity enables the leveraging of product variations by substitution of component variations

- Modularity helps contain change by enabling common components to be used within and across product lines

- Modularity facilitates decoupling technology development and product development, enabling concurrent and distributed development processes

---

1 In Dutch the project is called 'Gewild Wonen'. A literal English translation would be Sought-After Housing.

- Modularity enables the loose coupling of component designs and thereby creates loosely coupled knowledge domains.

Because of these properties, one of the most evident benefits of modularity is the ability to deliver mass-customized products and services. Mass-customization is an oxymoron combining the two contradictory notions of mass production and customization. Mass-market goods and services are uniquely tailored to the needs of the individuals who buy them. Mass-customization enables companies to customize individual offerings at little additional marginal costs. Without creating an overwhelming choice complexity or cost increase, it should provide customers exactly with what they want, for a value-for-money price level. Many diverse companies, like BMW, Dell Computer, Levi Strauss, Mattel and McGraw-Hill, are adopting mass-customiziation and often, the World Wide Web plays an important role in this strategy. Hart (1996:12) argues that "the mass-customization strategy is most effective for those firms that have leveraged their agile manufacturing capability by developing an integrated companywide system that includes collaborative design processes (often using CAD or AI software), flexible production processes and learning relationships."

Although modularity and mass-customization are closely connected. One should note that modularity in itself is not the same as mass-customization. It is only one of many possible ways to achieve mass-customization. Business solutions like customer relationship management, advanced production and manufacturing lines, the use of information and communication technology and dedicated inbound and outbound logistics are other building blocks for effective mass-customization (Pine 1993, Piller 1998). In the following sections we will consider the benefits of modularity in general, but focus in particular on the relationship between modularity in three dimensions (product, process and supply chain) and mass-customization.

2.2 Product Modularity

Most products are very complex and ill-defined; within each product level, many components may be linked to form the next higher level. This is especially true for customized products, while consumers may add certain attributes and drop others, or they may combine the product with another product that had been generally regarded as distinct. Alternatively, a product that consumers had treated as an entity may be divided into a group of subproducts that consumers can arrange into various combinations according to their personal preferences (Langlois and Robertson 1992). Because of this, it has been very hard to come up with an undisputed measurement of the modularity of a product, while each product is in fact different from the other. This may, for instance, lead to numerous (nested) levels of modularity within a single product.

Henderson & Clark (1990) distinguish between the components of a product and the ways they are integrated. The latter is defined as the product architecture; it is the scheme by which the functional elements of the physical product are arranged into physical chunks and by which the chunks interact (Ulrich & Eppinger 1995). The *functional elements* of a product are the individual operations and transformations that contribute to the overall performance of the product. The *physical elements* of a product are the parts, components and subassemblies that ultimately implement the products's functions. According to Ullrich and Eppinger (1995), a product architecture is fully modular if a 1:1 mapping exists between functions and physical components and de-coupled physical interfaces between interacting components. This means that a change made to one component does not require a change to other components for a correct functioning of the total product. A modular architecture has the following two properties:

• Chunks (modules) implement one or a few functional elements in their entirety

• The interactions between chunks are well defined and are generally fundamental to the primary functions of the product.

Modularity is a relative property of a product architecture. Products are rarely strictly modular or its opposite, integral. Rather, we can say that they exhibit either more or less modularity than a comparative product (Ulrich & Eppinger 1995).

Fine (1998) summarizes the, in his opinion, most essential differences between a modular and an integral product architecture. A modular architecture features separation among a system's constituent parts, whereby:

• Components are interchangeable

• Component interfaces are standardized

• System failures can be localized

In contrast, an integral product architecture might feature, for example:

• Components that perform many functions

• Components that are in close proximity or close spatial relationship

• Components that are tightly synchronized

In their most recent work, Baldwin & Clark (2000) elaborate further on modular design of products. They define a module as 'a unit whose structural elements are powerfully connected among themselves and relatively weakly connected to elements in other units. Clearly there are degrees of connection, thus there are graduations of modularity.' In other words, modules are units in a larger system that are structurally independent of one another, but work together. Baldwin & Clark state that after some analysis, they had to conclude that it is difficult to base a definition of modularity on functions, like Ulrich's (1995: 422), which are inherently manifold and nonstationary.

We however, follow the line of reasoning of Ulrich (1995). The mapping between functions and elements is essential: a 1:1 mapping denotes modularity, while a more complex mapping indicates integrality. Concepts like granularity (O'Grady 1999) and module attributes (Langlois & Robertson 1992) are in line with the function perspective. Further, the interactions (also called interfaces) between the modules are of great significance as well, together with the interdependencies between the modules and the connectedness within the modules. Standardized module interfaces enable the easy separation and recombination of modules.

2.3 Process Modularity

A process can be defined as 'a set of interrelated tasks to achieve a certain goal' (Davenport 1992). Business processes in particular refer to the technologies and procedures used to produce or deliver products or services within an organization (Boynton & Victor 1991).

Feitzinger & Lee (1997) argue that breaking down the production process into independent subprocesses provides companies with the kind of flexibility that effective mass customization requires. Such an approach is based on three principles: process postponement, process resequencing and process standardization. The three features closely resemble earlier-mentioned features of modular systems and products. The distinctiveness of process elements, their loose coupling and ease of switching and recombination indeed are apparent. Fine (1998) further elaborates on the coupling between process modules and introduces two dimensions in which the coupling between the modules can be defined: time and place. When the coupling between the process components decreases in time and/or place, the process becomes more modular. When certain process components are executed at a different moment or location, decoupled from the other process components, then process modularity increases. An example is the prefabrication of certain product components or the dispersion of teams to different locations.

To reach its cost and speed objectives, the auto industry is also moving to modular production. Parts from thousands of tier I and II manufacturers are being consolidated into large subassemblies by subcontractors, leaving just the final steps of assembling the car in the factory (Van Hoek & Weken 1998). To save time and money, such "modules" could eventually be put together "in-transit", thus avoiding costly physical facilities owned by the automakers. The retail dealer potentially may become the place of final assembly, moving manufacturing as close to the final customer as possible (Oliver 2000).

We observe that in itself, a process, like a production process or a distribution process, can become more or less modular too. When the coupling between the different process components decreases and the components become more distinct from each other, the process in itself becomes more modular. In line with this, Sanchez (1997) elaborates on previously-given definition of architecture by Henderson & Clark (1990) and extends it to the level of the business process. In his opinion, process architectures are conceptually analogous to product architectures. A modular approach intentionally tries to create a product or process design that permits the "substitution" of different versions of functional components for the purpose of creating product or process variations with different functionalities or performance levels (Sanchez 1997). When this analogy is indeed true and valid, previously defined features and properties of modular systems and products should be applicable to processes as well.

2.4   Supply Chain Modularity

Another type of modularity is only recently coming into attention, which is the modularity of supply chains (Venkatraman & Henderson 1998, Fine 1998, van Hoek & Weken 1998). The idea is that not only products and production processes can be modularized, but modularity can be applied in a supply chain setting as well. In analogy with modular products and modular processes, modular supply chains permit the "substitution" of different versions of functional components for the purpose of creating supply chain variations with different functionalities or performance levels[2].

---

[2] In this paper, processes refer to the technologies and procedures used to produce or deliver products or services within an organization (Boynton & Victor 1991), while a supply chain refers to the choice and design of the network surrounding an organization. The former says something about *how* the product is made or manufactured, while the latter refers to *who* does it.

As mentioned earlier, the opposite of modular is integral. Aldrich (1978) states that integration (of networks) is an ill-defined concept, making operationalization difficult and interpretation of outcomes confusing. Integration is generally considered to focus on issues of both interconnectedness among provider agencies and the extent to which provider agencies are integrated and coordinated through a central authority. Provan & Millward (1995) operationalize integration in line with the general network structure concepts of density (cohesion in a graph) and overall centralization (organization around particular focal points). Density is simply a measure of the extent to which all network organizations are interconnected. Centralization refers to the power and control structure of the network, or whether network links and activities are organized around any particular one or small group of organizations. According to Provan & Millward (1995) a high density and centralization refers to a high degree of integration. The opposite, low density and de-centralization, should therefore indicate a modular network.

The concept of density and connectedness is also used by Fine (1998), who introduces the proximity of elements to determine the degree of modularity (or integrality) of a supply chain. He argues that an integral supply chain architecture features close *proximity* among its elements. Proximity is measured along four dimensions: geographic (physical distance), organizational (ownership, managerial control and interpersonal and interteam dependencies), cultural (language, business mores, ethical standards and laws, among other things) and electronic (e-mail, EDI, intranets, video conferencing etc). A modular supply chain exhibits low proximity along most or all of the dimensions listed above. For instance, modular supply chains tend to feature multiple, interchangeable suppliers for key components (Fine 1998).

Subsequently, one can ask which type of coordination mechanism is most effective in each situation. Sanchez & Mahoney (1996) offer an answer by stating that in a loosely coupled system each participating component can function autonomously and concurrently under the embedded coordination of a modular product architecture. This corresponds closely to Daft and Lewin's (1993) notion of a modular organization, that continuously changes and solves problems through interconnected coordinated self-organizing processes. Fine (1998) also sees the analogy between modularity and a coordination strategy. 'Real competitive advantage could be made by individual organizations that know best how to design and coordinate their supply chain, i.e. how to efficiently source supply chain modules. The ultimate core competency is the competency of deciding which capabilities (modules) to make and which to buy, which are core and which are peripheral.'

2.5   Three-Dimensional Modularity

We saw in previous sections that different perspectives and definitions of modularity (and integrality) exist. Despite this dispute, we were able to unveil a number of so-called key characteristics and features of modular systems in general, and modular products, process and supply chains in particular. The following features were distinguished that are of importance to determine the degree of modularity:

- Distinctiveness of components

- Loose coupling between modules; tight coupling within modules

- Clarity of mapping between functions and components

- Standardization of interfaces

When all of these features are present, this should lead to easy separation and recombination of the components, just as the possibility to mix-and-match the components to come up with different varieties and structures.

While we now are able to determine the degree of modularity of a 'business system' (being either a product, a process or a supply chain), the next step is to consider the benefits of actually using a modular approach as opposite to a more integral one, especially when organizations are pursuing a mass-customization strategy. In particular, we want to investigate whether a concurrent design of all dimensions (product, process *and* supply chain) is preferrable above a more asynchronous approach where all three dimensions are designed independently of each other. We will commence with a theoretical discussion on this topic.

Fine (1998) claims that product, process and supply chain architectures tend to be aligned along the integrality-modularity spectrum. That is, integral products tend to be developed and built by integral processes and supply chains, whereas modular products tend to be designed and built by modular processes and supply chains. They tend to be mutually reinforcing and conducive to each other. The concurrency between processes and supply chains is also addressed by Feitzinger & Lee (1997) who state that manufacturing processes should be designed such that they, too, consist of independent modules that can be moved or rearranged easily to support different distribution-network designs.

This topic is closely related to the field of concurrent engineering. Concurrent engineering, according to the Institute for Defense Analysis (Handfield, 1994: 385), refers to "a systematic approach to the integrated concurrent design of products and related processes including manufacture and support". With concurrent engineering all activities required to bring a product to market-marketing, design, engineering, and manufacturing - are jointly managed to work in parallel, in sharp contrast with the traditional "throwing it over the wall" approach. That is, a typical company, much like a medieval castle, constructed protective walls around certain groups, functions or departments, in effect keeping out people who did not belong. The concurrent process has been characterized by Nonaka & Takeuchi (1995) as the "rugby team" approach to design instead of the sequential "relay race" approach.

Where concurrent engineering only focuses on products and processes, we follow Fine's (1998) line of reasoning and add the dimension of the supply chain to the former two. It is hypothesized that a concurrent design in these three dimensions leads to better mass-customization results and performance than a more asynchronous approach. The main objective of this research module is to investigate whether and under which circumstances this hypothesis is indeed valid.

3 Case Study: Dwelling on Demand

3.1 Background

Previous hypothesis is validated within the Dutch house building industry, which is currently trying to mass-customize its products. The housing industry in the Netherlands is a very traditionally organized industry with a low degree of industrialization and a large governmental influence. In most situations the customer is paying a large amount of money - a house is for most people the largest spending in their life - for a product they have no real saying in. Only during the last phase of most house building projects, the customer gets involved in choosing the colors of the tiles in the bathroom or he may select a kitchen from a showroom, in the best case.

At the moment the industry is more and more looking for ways to increase the influence of the customer in the design of their own house, without increasing the price too much and losing the advantages of serial, project-wise, production. Since the beginning of the 80's the property developer has become the main order placer in the Dutch house building industry. Formerly, housing cooperations functioned as professional order placers. The professional order placers are frequently involved in building projects and are experienced with the roles played by the building partners. Professional order placers could be own lodgers (like banks, governmental organizations or libraries), investors, social house builders or property developers. The other type of order placer is the incidental order placer that only places an order once in every five years at most. These order placers normally have little knowledge about the fulfillment of a building project. They need to insource the required knowledge from somewhere to make their project successful. Obviously, people need good guidance during this process.

The last couple of years, private, incidental order placementship (also called self-building) is strongly gaining ground[3]. It can be defined as 'building and developing houses in cooperation with the customer, where the customer may influence the architecture, the volume and lay-out of the house (within his own financial constraints)'. Many different types of self-building exist in practice. The first is traditional self-building, where the private person buys a plot and develops the house, possibly in cooperation with an architect and/or contractor. Other examples are collective order placementship and catalogue building (where the buyer searches in a catalogue the required house or composes the house with the help of examples).

A special type is customer-oriented property development. This is a fairly new development of which the Dwelling on Demand project is the main representative. It is introduced as a new, intermediate form between the existing models of self-building, where the advantages of the other types (both high customer influence as well as low costs and higher efficiency) are preserved. This new type may be described as the quintessential example of mass-customization in the Dutch housing industry.

3.2 Project Description

To explore the organizational and technical difficulties belonging to more mass-customization in the housing industry, the city of Almere, a 25 year old town in one of the Dutch polders, decided to initiate an experimental project in early 1999. They called it Dwelling on Demand. It was decided to invite fifteen professional property developers to come up with architectural designs, based on the theme. The project in its entirety had to display the image of housing in the 21st century

---

[3] The percentage of self-building in the annual production of houses has, since 1983, slightly increased from 12% to approx. 17% in 1998. The Dutch government wants this percentage to become 33% in 2005. A large variation can be found in this percentage depending on the exact location in the Netherlands. Nowadays, urban highly dense areas only reach 6% self-building, where rural areas already consist of 30% self-building.

according to the initiators. The objective was to design and build houses with such a variety that on the one hand no house would look the same, while on the other hand the advantages of project-wise building were preserved. The project is one of the first consumer-oriented property development projects in the Netherlands.

The initiators of the project chose the following plan of approach. First, the fifteen architects received the assignment to design just a base, a core, of the house with the necessary utilitarian and sanitary provisions. This design should not be a complete house, but just the start of a house. Subsequently, the architect was supposed to deliver a catalogue of components to complete the house. Using the components the house could be assembled, in much the same way a car is assembled. The style, material and size had to be determined by the (future) owner of the house. The components chosen from the catalogue of styles could be assembled on the building site. The exact position of the house on the building plot should be a free choice, as long as the privacy of the adjoining dwellings was not hampered. The appearance of the dwellings was bounded by only a minimum of rules. Only the building height was limited to three floors. Flat roofs or topped roofs, both were allowed. Also the shape of the roof was free, as well as its slope and its direction.

3.3   Three-Dimensional Modularity

The degree of product modularity employed, largely differed among the 15 different sub-projects. We distinguished between three product levels: exterior, interior and accessories. The designs with the highest possible degree of customization were the ones that employed modularity on the level of the exterior of the house. In these cases, customers were not just offered a number of standard house types to choose from. Instead, they now got the opportunity to really design the shape and size of their own house, by selecting e.g. roof, front, floor or additional volume modules.

Even in these designs it remained the architect who decided how much freedom a buyer would get. Each sub-project consisted of 40 houses on average and each architect somehow wanted to make sure that the total image of the individual houses in the sub-project remained coherent - there had to be some repetition and similarity in the houses. Architects who chose repetition on a generic, higher level (exterior of the house) often did this for building efficiency reasons, with limited customer freedom as a consequence. Repetition on a more detailed, subtle level (e.g. separation walls, window frames) often led to more freedom and more variety in designs.

Real modular housing kits, comparable with Lego-systems or the *modular homes* often built in the US, have not been designed. A house is not suited for that; an attractive architectonic image and the use of a modular system conflict with each other, according to the majority of the architects. Furthermore, the mapping between function and modules was not fully modular like described earlier when discussing modular products. Neither was the standardization of interfaces – the scheme by which the modules are connected. The Dutch building industry has not developed such standardization yet. As a consequence, the architects had to design every specific interface in detail.

Even for regular housing projects in the house building industry, a strong linkage exists between the design of the product, the design of the production processes and the accompanying supply chain of participating organizations. After the initiative phase, the architect comes up with the basic shape of his design; from an urban design point of view she designs houses with a rigid shape: connected, piled up, semi-detached or in blocks. Already when choosing the shape, the architect needs to know how the houses will be built in a later stage. For instance, when the hull is made of concrete, it needs to be assured that the hoisting crane can reach the building areato move the formwork elements during the construction. When the houses are built in a densely built area and no building site is available, one will often choose for building with prefabricated elements for the hull, thus avoiding the use of large equipment. In addition, a building company constantly has to consider whether a design can be produced effectively with a certain building method. The fact whether the building company involved in the project is specialized in only one building method or can work with different methods is also of importance in that case.

In the Dwelling on Demand project this link was further strengthened. It was found that the designs, which employed the highest degree of product modularity and thus, highest degree of customer influence, had to make use of the most flexible and modular production processes. Many of the building companies were forced to switch from a building method in which they had specialized for years (e.g. concrete building) to another, more modular method (e.g. woodframe building). In general one was looking for building methods that kept the final decision of the customer open as long as possible; where it is possible to make late changes. The often-applied 'Open Building' method is such a method that further stresses the concurrency between both dimensions. With this method it is possible to produce different houses, while using the same hull. The hull is fixed, while the infill (inner walls, pipes and installations) can be carried out in different ways. This enables that within a fixed hull, mutual different floor plans and layouts can be made. However, it also means that the variable building elements need to be manufactured and assembled, such that this easy replacement is indeed possible: the elements need to be modular with standard interfaces and sizes.

With respect to the modularity of supply chains it was found that all organizations chose a very modular type of supply chain. In the building industry, the exact way the different stakeholders cooperate, how risks and responsibilities are divided and which roles are fulfilled by the stakeholders, is determined by the chosen organizational model. In the Dwelling on Demand project, the organizational model chosen by all organizations was the building team, consisting of the property developer, the architect, several specialists (advisors) and one or more executing building companies. The team collectively realizes the design, where each member brings in his own competencies, while design and execution responsibilities are explicitly separated. Compared to other supply chain models (e.g. traditional approach, general contracting, and brochure plan), the building team is a very modular form with clearly distinct elements and standard interfaces (contracts). However, even the building team could be subject to changes compared to regular building projects. We investigated whether the building team actually became more modular or not.

What we saw most of the times in the sub-projects was that whenever both product and process modularity were low, the supply chain (i.e. the structure of the building team) did not require much changes compared to normal. This was in these cases very succesful, at least when the customer's disposition to participate in the design was also low. The stakeholders in these projects could rely on well-proven organizational structures and efficient and relatively low-cost building methods.

Other projects demonstrated a high degree of modularity in all three dimensions. Such a modular, concurrent design proved to be more difficult to accomplish than previously described designs. In these cases, Information and Communication Technology proved to be really needed, customers require support and guidance and the entire network must be ready for this. It turned out that some sub-projects were better equiped for this than others. The successful ones made use of ICT to support their customers, they were really looking for multiple suppliers for their key components and used these suppliers' creativity and knowledge wherever they could. They thus managed to make the project profitable and make their customers, with a high disposition to participate in the design, satisfied. Other sub-projects however, did not manage this. Their customers were not satisfied and the project was loss-making. Probably they underestimated the complexity of the project and ended up in serious shortage of time. This subsequently limited customer freedom, which could have been much higher due to the modular product design.

Finally, we found two very ambitious projects where it was decided not to use a modular product design at all. These designs only consisted of the basics of a house (a small kitchen, a toilet and a small room) where people could live in during the building process; the remainder of the house had to be designed and built by the customer himself. These designs did not use choice options, questionnaires or variants; customers were offered almost complete freedom. In the end however, these projects failed to live up to their expectations. Probably both projects were too ambitious: both the network participants as the customers themselves were not ready for this concept. The customers' disposition to participate and their know-how were in both projects too low to justify such an ambitious design. The developers themselves were not ready for this either. ICT infrastructures were not in place, customer support could not be offered and so on.

It was concluded that, to keep these type of houses affordable and feasible for the customer, one surely needs modularly designed supply chains, which make it possible to configure a cut-to-fit supply chain for each individual house design. In this case, if the property developer wants to support the customer in keeping the costs low, he needs to have access to a very modular network of actors. For each individual customer he needs to be able to set up a temporary, loosely coupled network of organizations, willing to participate in the design of this house.

4   Conclusions

This paper makes three key contributions to research on modularity and mass-customization.

First, we identified the key role of three dimensions of modularity: product, process and supply chain modularity. The different aspects and opportunities of three-dimensional modularity have been analyzed in relation to successful mass-customization strategies. It was shown that the features of a modular design - distinctiveness of components, loose coupling between modules; tight coupling within modules, clarity of mapping between functions and components and standardization of interfaces – were applicable on all three dimensions.

Second, by means of a detailed analysis of an innovative mass-customization housing project, we validated the main proposition that a network of organizations will be more effective in pursuing a mass-customization strategy when all three dimensions are concurrently designed in a modular fashion. A concurrent design in all three dimensions often leads to better performances, while it assures a better fit between serving the customers' requirements and the organizational network structure and capabilities.

Third, it was found that, when mass-customizing products, the willingness and ability of the customer to participate in the design of the product, and thus the network, is a very important factor that should be taken into consideration. Furthermore, the more modular the design, the more ICT becomes an indispensable means to solve all complexity problems.

References

Aldrich, H.E. (1978): Centralization Versus Decentralization in the Design of Human Service Delivery Systems: A Response to Gouldner's Lament, in: Issues in Service Delivery in Human Service Organizations, (eds. R. Sarri and Y. Hasenfeld), New York: Columbia University Press, pp. 51-79.

Baldwin, C.Y., K.B. Clark (2000): Design Rules - Volume 1. The Power of Modularity, The MIT Press, Cambridge, Massachusetts.

Boynton, A.C., B. Victor (1991): Beyond Flexibility: Building and Managing the Dynamically Stable Organization, California Management Review, Fall 1991, pp. 53-66.

Daft, R.L., A.Y. Lewin (1993): Where are the theories of the "new" organizational forms? An editorial essay, Organization Science, 4, 4, pp. i-vi.

Davenport, T.H. (1992): Process Innovation, Harvard Business School Press, Boston.

Feitzinger, E., H.L. Lee (1997): Mass Customization at Hewlett-Packard: The Power of Postponement, Harvard Business Review, Jan.-Feb. 1997, pp. 116-121.

Fine, C.H. (1998): Clockspeed - Winning Industry Control in the Age of Temporary Advantage, Perseus Books, Reading, Massachusetts.

Henderson, R.M., L.B. Clark (1990): Architectural Innovation: The Reconfiguration of Existing Product Technologies and the Failure of Established Firms, Administrative Science Quarterly, 35, pp. 9-30.

Handfield, R. B. (1994): Effects of concurrent engineering on make-to-order products, IEEE Transactions on Engineering Management, 41, pp. 384-393.

Hart, C.W. (1996): Made to Order, Marketing Management, 5, 2, pp. 10-22.

van Hoek, R.I., H.A.M. Weken (1998): The Impact of Modular Production on the Dynamics of Supply Chains, The International Journal of Logistics Management, 9, 2, pp. 35-50.

Langlois, R.N., P.L. Robertson (1992), Networks and innovation in a modular system: Lessons from the microcomputer and stereo component industries, Research Policy, 21, pp. 297-313.

Nonaka, I. H. Takeuchi (1995), The Knowledge Creating Company, Oxford University Press , New York, NY.

O'Grady, P.J. (1999): The Age of Modularity: Using the new world of modular products to revolutionize your corporation, Adams and Steele Publishers, Iowa, USA.

Oliver, R.W. (2000): AutoIntelligence, Management Review, New York, Feb 2000.

Piller, F.T. (1998): Kundenindividuelle Massenproduktion: Die Wettbewerbsstrategie der Zukunft, Carl Hanser Verlag, München (in German).

Pine, B.J. (1993): Mass Customization: the new frontier in business competition, Harvard Business School Press, Boston, Massachusetts.

Provan, K.G., H.B. Milward (1995): A Preliminary Theory of Interorganizational Network Effectiveness: A Comparative Study of Four Community Mental Health Systems, Administrative Science Quarterly, 40, pp. 1-33.

Sanchez, R. (1997): Preparing for an Uncertain Future - Managing Organizations for Strategic Flexibility, International Studies of Management and Organization, 27, 2, pp. 71-94.

Sanchez, R. (1999): Modular architectures in the marketing process, Journal of Marketing, 63, Special Issue, pp. 92-111.

Sanchez, R., J.T. Mahoney (1996)_ Modularity, flexibility and knowledge management in product and organization design, Strategic Management Journal, 17, Winter special issue, pp. 63-76.

Ulrich, K.T. (1995): The role of product architecture in the manufacturing firm, Research Policy, 24, May 1995, pp. 419-440.

Ulrich, K.T., S.D. Eppinger (1995): Product design and development, McGraw-Hill, New York, NY.

Venkatraman, N., J.C. Henderson (1998): Real Strategies for Virtual Organizing, Sloan Management Review, Fall 1998, pp. 33-48.

Customization of Capital Goods – Implications for After Sales

Suomala, P., Sievänen, M. & Paranko, J.

Summary: Customization is nowadays often regarded as a dominant paradigm in many industries. However, the impacts of customization on different business perspectives are still inadequately reported. Especially in the case of long product life cycles, after sales is a very significant business domain. In addition, it is even claimed to be a very profitable one. A distinctive feature of after sales business is the high number of low volume product items. The question is, does customization actually produce more those kinds of items, and thus increase the costs and complexity of operations. A case study is conducted to investigate the implications of customization for the spare part business. In this study, the complete life of one product cycle up to the present is analysed. The main focus is on the change in the number of spare part items and the change in inventory value due to customization. A set of explanations is provided to describe the relationship between after sales and customization. Furthermore, a classification of customizations from the after sales point of view is presented in this paper. As a conclusion, the study suggests that the impacts of customization on the number of items and inventory value are minor.

Keywords: customization, after sales, product items

1 Connecting customization with after sales

Customization has been mentioned as one of the marketing practices of the 21st century. Small export firms have met the needs of their overseas markets with customized products and have gained larger market shares[1]. Better service and increased customer value provided by customization have boosted profits in many

---

[1] (Sriram and Sapienza 1991)

cases[2]. After a firm has customized a product or a service, it knows the customer's preferences and so next time it can tailor its business to fulfill the customer's needs better than competitors[3]. In this way the firm may positively influence the customer's loyalty. Some studies have identified customization as a means of improving customer satisfaction[4]. It is said that there is a growing demand for customized products and they are perceived as a status symbol[5]. Moreover, customers will pay a premium for goods and services which satisfy their unique needs[6].

Although the majority of the discussion concerning customization has been more or less marketing of the idea, some critical aspects have also been addressed. Radder and Louw note that customization does not necessarily fit in every company. They suggest that mass production is still a feasible policy in many cases[7]. To some extent writers have also sought compromises between customization and mass production[8].

Ulrich and Eppinger define customized products as follows: "Customized products are slight variations of standard configurations and are typically developed in response to a specific order by a customer."[9] A customized product can be seen as a generic product which is modified by customer needs, like a car with a list of optional extras. Or it can be seen as a special product which is made of standard modules combined in the way the customer wants, like a prefabricated house. A unique product, like a work of art or a prototype, is not a customized product even if it has been made to fulfill a customer order.

There are many different types of customization. An interesting classification is presented in Lampel and Mintzberg's article[10]. They define a manufacturing firm's value chain in four stages: design, fabrication, assembly, and distribution. Depending on the part of the value chain in which the customization is made, they have identified up to five different strategies:

---

[2] (Simon and Dolan 1998)
[3] (Computing Canada 1998)
[4] (Fitzgerald 1995, p. 44), (Zairi 2000)
[5] (Evarts 1999)
[6] (Lipe 1995)
[7] (Radder and Louw 1999)
[8] (Jiang 2000)
[9] (Ulrich and Eppinger 1995)
[10] (Lampel and Mintzberg 1996)

- *Pure Standardization.* No customization is made by the manufacturer or distributor. For example: the Model T Ford.

- *Segmented Standardization.* Customers are seen as a cluster of buyers and each cluster is seen as a whole. For example: different types of Coca-Cola.

- *Customized Standardization.* The assembly is customized. A customer can select from a list of options which are made from standard components. This can also be called modularization or configuration. A car is a typical example.

- *Tailored Customization.* A generic prototype is presented to a customer and then tailored according to the customer's individual needs. Customization is done in distribution, assembly, and fabrication. A tailor-made suit is an example.

- *Pure Customization.* A customer influences the entire production process even at the design stage. In large-scale construction projects like paper mills or nuclear power plants everything can be customized; projects like the Olympic Games or NASA's Apollo are extreme examples. In some cases this type of customization does not fulfill the previous definition and these products can be considered as completely new products because there is no standard configuration.

Lampel and Mintzberg perceive customization as a particular strategy. Coates and Wolff approach it as a manufacturing practice[11]. They define customization as soft when a customer does not actively intervene in the manufacturing process. Soft customization can be seen as a synonym for customized standardization. Their definition of hard customization is very close to tailored customization.

In the 1990's companies started to use mass customization as a manufacturing strategy to supply unique goods and services to their customers. Gilmore and Pine have identified four distinct approaches to mass customization: collaborative, adaptive, cosmetic, and transparent[12]. In collaborative customization the customer takes part in the design process, but manufacturing and assembly processes may be standard. Distribution is customized. In adaptive customization, processes are typically standard, but customers can easily customize the product on their own. A software configuration is a typical example. Cosmetic customization is very close to customized standardization. Customers use the product in same way but they want it to be presented in a different way, e.g. in a different size of package. Transparent customization is almost the opposite to cosmetic customization.

---

[11] (Coates and Wolff 1995)
[12] (Gilmore and Pine 1997)

Typically the package and the delivery process are the same, but the product itself may be customized. Thus, the customer's preferences are considered in the design process, but everything else is done in the standard way. Furthermore, the customer may not even know how the actual product is customized. In all these mass customization approaches the manufacturing is quite standard.

Spring suggests that the customization may have different roles in an organization, which furthermore may be partly overlapping with each other. Four different motives for customization have been identified[13]:

- Entry barrier
- Vehicle for learning
- Symbol to industry
- Profit-taker

Two issues can be added to the previous list. The first is increasing the market share and the second is conduction of market- or customer-driven R&D. Obviously R&D and learning are closely connected with each other. In general, customers demanding customization should not be perceived as a nuisance, but rather as an opportunity to expand the business.[14]

One of the fundamental questions regarding customization is how customization affects the customer-perceived value. Lapierre has addressed the question in the capital goods market. According to his study, the availability of customized products as such was not perceived as more valuable than the availability of alternative solutions in general. From the customer point of view, value creation was more based on service-related factors such as flexibility and responsiveness than on product-related features.[15]

Manufacturing operations have typically had a mass production mentality and current business processes were developed to support this mass production mind-set. These processes do not meet today's market demands. Companies have to view customization as another customer demand that challenges the ability to maintain the cost, quality, and speed of operation. Companies have to find new ways of thinking and design business processes to serve customers who demand specialization.[16]

---

[13] (Spring and Dalrymple 2000, p. 462-463)
[14] (Gilmore 1993, p. 25)
[15] (Lapierre 2000)
[16] (Gilmore 1993)

When a company starts customization, typically its variety of products increases and batch sizes become smaller. In manufacturing this means typically more set-ups and changeovers. Ancillary costs are incurred every time when a machine is set up or changed over. Thus, there is a challenge to reduce set-up expenses in order to be able to do more set-ups. That is because customers demand more customized products.[17]

There are many ideas on how to respond to these challenges. Quinn and Paquette talk about dividing processes into the smallest possible core units and then managing these micro-units[18]. Gilmore argues that companies have to make the following three strategic changes[19]. Firstly, change the time orientation so that time flows from the future back to the present. Secondly, change the work orientation from functional to process. And finally, use new tools which emphasize the underlying structure of processes. What is needed is systematic thinking to generate alternative concepts.

There is not enough detailed research evidence available concerning the implications of customization for different business processes and their costs in particular. For instance, occasionally customization is too easily associated with small volumes. Although there is typically a positive correlation between small volumes and great variety, it does not mean that customization always leads to a direct increase in the variety of all items.[20]

Mughal and Osborne point out that the number of items is a significant cost driver in all types of organizations. Therefore, standardization should be promoted at least in those items or features that are not directly visible or noticeable to a customer. From the profitability point of view, product engineering should be completed so that once the product market launch has happened, the original product specification is no more revisited and altered. The features should be engineered for the whole life-cycle.[21] The principles presented by Mughal and Osborne are basically prerequisites for efficient configuration, but they do not have any concrete evidence regarding the economical benefits achieved by those principles.

Åhlström and Westbrook have surveyed the implications of customization utilizing a questionnaire. Most of the respondents (companies) carried out the customization in the assembly-phase: individual products were created from standard modules

---

[17] (Darlington 1999)

[18] (Quinn and Paquette 1990)

[19] (Gilmore 1993)

[20] (Spring and Dalrymple 2000, p. 445)

[21] (Mughal and Osborne 1995)

that were combined in a unique way. The benefits of the customization were perceived quite homogeneously: customer satisfaction and market share increases. The disadvantages of the customization were associated with increasing material and manufacturing costs. Furthermore, the understanding of customers' needs and the management of the whole supply chain were perceived as the most important challenges of customization.[22]

The spare part business is assumed to be a very profitable business area in many cases. There is no reliable data available, but it is commonly believed that spare parts create one third of the net sales and two thirds of the profit. Depending on the customer's requirements, the company has to choose different service and spare part strategies. The capital goods manufacturer typically chooses a rapid response service strategy[23]. When breakdown occurs, the opportunity costs are generally high so the downtime has to be minimized. This means that the company has to keep certain critical spare parts available at all times.

Because customization means differentiated products, it typically increases the variety of products. Moreover, it may increase the number of low volume items. When we are talking about spare parts, this means new items to be kept in stock. However, only a part of the items of the final product are classified as spare parts. In the machine and metal product industry it is common that approximately 10% of all items are classified as spare parts. Furthermore, quite often only half of the spare parts are kept in stock[24]. Spare parts are typically low volume items. For example in the Royal Dutch Air Force, over 90 % of the items are needed ten times or less per year[25]. The control of a spare part inventory is a complex matter, not only because of low volumes per item, but also because the demand is very fluctuating. Moreover, long life cycles and a long guaranteed period of spare part supply (>20 years) make inventory control even more demanding.

Based on a German study, the average spare part inventory of a metal industry company with sales on capital goods less than € 50 million was € 4.2 million[26]. Furthermore, the carrying cost is between 20% and 40% of the inventory value[27]. Carrying costs include the cost of space, insurance, obsolescence, material handling, and the capital costs. Because of the high costs and increasing inventory

---

[22] (Åhlström and Westbrook 1999)

[23] (Lele 1997)

[24] (Fortuin and Martin 1999, p. 959)

[25] (Fortuin and Martin 1999, p. 955)

[26] (Pfohl and Ester 1999)

[27] (Deierlein 1998; Sandvig and Allaire 1998; Transportation & Distribution 1999)

value, the inventory management gets more attention. In addition, stock-out costs can be high in terms of lost customers.

2 Research design

2.1 Objectives

The objective of the study is to point out the consequences of product customization on after sales operations. By increasing the understanding regarding customization and the nature of the spare part business, the study is to provide a set of explanations to assess the relationship between product customization and the costs of after sales business. The scope of this study is to evaluate the effects of customization on the number of after sales product items. In addition, the impact of product customization on the spare part inventory value is calculated.

A case study approach is adopted to answer the following four research questions:

1. What is the effect of customization on the number of product items in after sales?

2. What is the effect of customization on the spare part inventory value?

3. What is the technical nature of the customizations and how are the customizations actually carried out?

4. How can different customizations be categorized from the after sales point of view?

According to Yin[28], a case study is an empirical inquiry that investigates a phenomenon within its real-life context, when the boundaries between phenomenon and context are not very evident. Furthermore, "how" and "why" type of research questions, in addition to "what" questions, are likely to favor the use of case studies. All these conditions apply to this research. The first two questions are descriptive (or exploratory) by nature. The objective is to discover the fait accompli regarding the relationship between customization and after sales in a case product. Furthermore, the following two questions provide relevant explanations for the discovered facts. Thus, they are investigated for explanatory purposes.

---

[28] (Yin 1994, p. 13)

The first question has two perspectives. Firstly, it has to be assessed whether customization increases the number of items or not. Secondly, the extent of possible increase needs to be conditioned against relevant reference values. The second question entails answering if customization has implications on the value of the spare part inventory. If that is the case, the effects have to be, for example, compared with the total value of inventory. Answering the third question provides information that helps to understand the events noted in the first two questions. The last question is about constructing a framework for the connection between customization and after sales business.

2.2 Research propositions

The case study relies on a single case product of one company. The rationale for this is that the product represents a critical case. The case product is seen, by the company representatives, as a typical example of a substantially customized product. According to Yin[29], when a theory has specified a clear set of propositions, as well as the circumstances within which the propositions are believed to be true, the single case can be used to determine whether a theory's propositions are correct, or whether some alternative set of explanations might be more relevant.

According to a proposition set by the company representatives before the study, product customization would have negative impacts on the spare part business. These alledged impacts can be divided into two categories: 1) the increase of the number of spare part items and the increase of the inventory related to that, and 2) the increased complexity of operations and time consumption of employees.

The underlying proposition of the study is that customization indeed has effects on after sales[30]. However, customization cannot be considered a uniform phenomenon. Instead, different customizations have different effects depending at least on the following:

- Target of customization

- Technical nature of customization

- Selling and purchasing volume of the customized item

- Degree of modularity of product structures or degree of collectively used parts

---

[29] (Yin 1994, p. 38)
[30] (See e.g. Suomala, Sievänen et al. 2000)

It has to be emphasized that, in this paper, the studied effects of customization do not cover the whole company or supply chain. If customization has a negative impact on after sales, it should not be interpreted so that customization has a negative impact on the business, as a whole.

For managerial purposes, a categorization of customizations based on their effects on after sales would be beneficial. After sales provides one aspect when evaluating and measuring the quality or success of an R&D decision. Within the scope of this study, such a categorization is presented in this paper.

2.3 Collection of data

The collection of data was made during the year 1999. It consisted of mainly three phases. The first stage was carried out by the researchers only, while the rest were conducted in cooperation with the company representatives.

The first phase involved clarifying the individual customizations made during the case product's life cycle. For this purpose, all the sales documents regarding the product were studied by the researchers. The main goal was to identify the orders that were outside the standard configuration. As a result of this phase, a database concerning customized products was created. It consisted of, in addition to customizations, product ID, product type[31], responsible engineer, customer, order date, and delivery date. The database was a foundation for the further stages of the study.

The second phase consisted of interviews of responsible engineers. Altogether four R&D engineers were interviewed. Firstly, each customization was studied separately and the engineers were asked to comment the technical nature of the customizations in detail. Secondly, the engineers' opinions about the effects of the customizations on after sales were collected. All the questions in the second and third phase were open-ended.

The third phase was conducted in cooperation with the after sales department. Firstly, two after sales managers were interviewed to find out the general implications of the customizations that had been made. The respondents were asked

---

[31] Product type is a code used by the company to describe the degree of standardity of the product. It is a kind of universal indicator that is not determined on the basis of a single perspective of the business. Code A means that the product contains only standard options. B1/B2 represent situations where there are some minor/major customizations made, respectively. Product code C indicates that the product is completely non-standard.

to assess the effects on two levels: 1) product order (the whole product with all the customizations made to it), and 2) individual customization. While the focus was on the level of individual customization, the two-level questions were to ensure the reliability of the answers. The scale utilized in this close-ended section is demonstrated in Table 1. The question was whether a customization creates a new item to be added to after sales or not. Further, the respondents were asked if a customization creates a new item that is: a) critical b) kept in regional inventory c) kept in local inventory. Critical means that the item is critical in terms of the customer's operations. Therefore, a quick delivery of those items has to be guaranteed either by keeping the item in stock or by arranging a deal with a subcontractor. If an item is included in both regional and local inventories, its impact on inventory value is higher than that of an item kept only in regional stock. All the relevant combinations are indicated utilizing codes from 0 to 5.

Table 1: The classification of customized items

Code	New items	Critical	In inventory	
			Regional	Local
1	X	X	X	X
2	X	X	X	
3	X	X		
4	X		X	
5	X			
(0)				

Secondly, all the customizations that would cause new items to be kept in stock were analyzed in detail. The product structures and stock values of the items were clarified through interviews. As a result, the total impact of customizations on inventory value was calculated.

3   The case

3.1 Introduction

The case company is a large metal industry firm operating worldwide. It designs, manufactures, and sells a wide range of products for industrial customers. The average unit price of an end product of the company is in the range of one million USD and the typical length of the product life cycle is over 20 years. Although the company can offer a variety of standard products with several standard options, customization is a typical phenomenon in this industrial environment. In fact, most of the sold products are actually unique in nature. Customization is thus defined as not only selecting a collection of standard options for the end product, but also selling and designing some additional unique features or attributes into the standard product. A customized product may, for instance, have a custom color or different power source than the standard product.

The case product is a typical representative of the company's product range. The physical dimensions are large, and the product weighs 59 tons. The product was introduced in 1994. The selling volume is typically 10 - 20 per year, the total machine population at the moment being ca. 50. Roughly one half of the sold products are somewhat customized.

3.2 Empirical evidence

A total of 42 product orders were investigated. The company had categorized 11 of them as being class A products, 16 being class B1, 13 products being B2 and only two products fell into class C. The total number of customizations was 196, while the number of different customizations was 122. Thus, a similar customization was repeated less than two times, on the average.

Table 2 summarizes the frequencies of customizations made for different product classes. A great number of type 0 customizations (109 out of 196) indicates that most of the customizations do not have any direct effect on after sales.

Table 2: Relationship between the impact[32] of customization and product class

Impact on AS	Customizations per product class/ (average number per product)				
	A (n=11)	B1 (n=16)	B2 (n=13)	C (n=2)	SUM
0 (nothing)	12 (1.1)	44 (2.8)	40 (3.1)	13 (6.5)	109
1 (all stocks)	0 (0.0)	0 (0.0)	0 (0.0)	0 (0.0)	0

[32] See Table 1 for interpreting the impact coding

2 (critical, regional stock)	0 (0.0)	0 (0.0)	1 (0.1)	2 (1.0)	3
3 (critical)	3 (0.3)	3 (0.2)	1 (0.1)	1 (0.5)	8
4 (regional stock)	1 (0.1)	4 (0.3)	4 (0.3)	2 (1.0)	11
5 (new items)	10 (0.9)	20 (1.3)	29 (2.2)	6 (3.0)	65
SUM	26	71	75	24	196

The number of customizations that actually add new items to after sales (types 1 - 5) is 87. A new item is not, however, necessarily kept in stock. In most cases, the creation of a new item means that the information systems and documentation have to be updated, but any direct action besides that is not needed. Indeed, only 14 customizations (types 2 and 4 together) would lead to the creation of new stock item(s). These fourteen customizations were taken under further investigation to clarify the impact of customization on stock value.

It is worth noticing that class A products caused a minimal amount of new stock items. Only one out of a total of 26 customizations had an impact on the inventory. Roughly half of the cases (12) did not have any effect on after sales. The profiles of classes B1 and B2 do not significantly differ from each other. The majority of customizations concerning both classes did not even add new items to information systems. Nine customizations led to an increase in the inventory value by producing new stock items. One third of the customizations (49 out of 146) created new items that are not in stock.

The number of investigated product orders in class C is essentially smaller than in other classes. However, the table shows that the implications for after sales regarding these orders are clearly more notable. On average, each order consisted of two customizations (one of type 2 and one of type 4) having an impact on the inventory. In addition, each order had created three new after sales items (that are not kept in stock), on average.

One interesting perspective regarding the effects of customization on after sales is to examine the relationship between the technical nature of customization and the type of impact. It is assumed that the extent and the type of impact would somehow depend on the technical nature of customization. In this study, all the customizations have been categorized firstly based on whether they are, by nature, additions to, alterations, or removals from the standard configuration. Secondly, based on the target of customization, they are classified into the following classes: electric, hydraulic, subassembly and mechanic.

The electric and hydraulic types of customizations concern an electrical/ electronical or hydraulic part of the product, respectively. Subassembly refers to a target of customization that requires a large subassembly of different kinds of items. The mechanic class consists of customizations that can be made, for instance, by welding, reshaping, painting or by adding a fairly simple mechanical part.

The first column of Tables 3, 4, and 5 shows the impact of customization on after sales based on the categorization made in Table 1. The second column indicates the technical nature of customization. The third one illustrates how many different kinds of customizations have been made that match these definitions. The last four columns show the degree of customization made in each product class.

Table 3: The impacts[33], types and targets of customizations when something is added to the standard configuration

Impact on AS	Technical type	Different customizations	Target product class (number of)				SUM
			A	B1	B2	C	
0	add;electric	3	1	1	1		3
3	add;electric	1	3	3			6
5	add;electric	11	5	2	9	3	19
0	add;hydraulic	2		3	3		6
3	add;hydraulic	1				1	1
4	add;hydraulic	1			1		1
5	add;hydraulic	5	1	2	3		6
2	add;subassembly	1				1	1
4	add;subassembly	1			1		1
5	add;subassembly	1	1				1
0	add;mechanic	20	5	13	9	3	30
4	add;mechanic	2	1	1		1	3
5	add;mechanic	11	1	8	8	1	18
SUM			18	33	35	10	96

---

[33] See Table 1

The total number of addition (96) and alteration (93) type of customizations is almost equal. Regarding both these customization types, most of the customizations are mechanical by nature. There are 51 mechanical additions and 75 alterations, altogether. However, only some of the mechanical customizations actually led to an increase in the number of stock items: the number of those in Table 3 is three (the second-lowest line) and in Table 4 the number is two.

Table 4: The impacts, types, and targets of customizations when something is altered compared to the standard configuration

Impact on AS	Technical type	Different customizations	Target product class (number of)				SUM
			A	B1	B2	C	
0	alt;electric	2		3		1	4
0	alt;hydraulic	1		1			1
3	alt;hydraulic	1			1		1
5	alt;hydraulic	2			2		2
0	alt;subassembly	2	1	1	1	1	4
2	alt;subassembly	2			1	1	2
4	alt;subassembly	1		3	1		4
0	alt;mechanic	38	4	21	22	7	54
4	alt;mechanic	1			1	1	2
5	alt;mechanic	9	2	8	7	2	19
SUM			7	37	36	13	93

Thus, the question remains: Which type of customizations would cause the most implications for after sales? Based on the information presented in these tables, it seems clear that subassemblies had the strongest effect on the spare part business. Two out of three subassemblies in Table 3 had an impact on the inventory. Moreover, six out of ten subassemblies in Table 4 had the same effect. Hydraulic and electric customization had minimal effects on after sales. This is mostly due to the fact that those components are provided by subcontractors. Thus, the after sales implications concern mainly them.

245

Customizations that were conducted by removing something from the standard configuration are presented in Table 5. These kinds of changes did not have any implications for after sales.

Table 5: The impacts, types and targets of customizations when something is removed from the standard configuration

Impact on AS	Technical type	Different customizations	Target product class (number of)				SUM
			A	B1	B2	C	
0	remove	3			3		3
SUM			0	0	3	0	3

The effects of customizations on the inventory value were calculated on the basis of the number of type 2 and 4 customizations. The conclusion was that all the customizations made for the case product resulted in five new stock items for the regional inventory. The total increase of inventory value due to this was 36,000 FIM (~6,000 Euros). The increase is regarded as minimal compared, for example, to the total value of the inventory which is 29 million FIM (~4,8 million Euros).

4 Discussion

The spare part business is an interesting and economically significant business area in many industries today. It is often even claimed to be the most profitable function of a corporation.

The intent of this study was to point out the implications of product customization for after sales. A case study approach was adopted to investigate the effects of customization on the number of spare part items and, further, on the inventory value. Prior to the study, the case company representatives had a strong belief that customization has a negative impact on after sales. Therefore, the study was not only to explore the effects of customization but also to find causes for these effects by categorizing different customizations and product orders.

This study suggests that the direct implications of customization for after sales are small. In general, customization tends to increase the number of items that are stored in information systems. However, only a minority of items is kept in stock. Furthermore, a great deal of customizations can be realized by altering or adding a fairly simple mechanical structure to the product. In that case, spare part support is rarely needed after sales. The more complicated subassembly is customized, the more effects customization has on after sales. It is interesting to notice that the classification system of product orders (A, B1/2, C) also supports the evaluation of implications of customization for after sales quite well.

The main contribution of this study to the others conducted in this field is the combination of after sales and customization. Customization has been widely researched from the product design, marketing, or manufacturing perspective. However, enough attention has not been paid to after sales and the spare part business. In this context, this study has contributed to the technical classification of customizations and to the categorization of customizations concerning its impacts on the number of product items and inventory. It is believed that the categorizations have been made detailed enough to have not only scientific but also managerial relevance.

The results of the study have been somewhat unexpected for the company representatives. The impacts of customization were assumed to be far more negative. The main reasons for the conflict between the expectations and the findings could be: the single case that cannot describe the cumulative impacts of all customized products, the relatively narrow focus[34] of the study (all kinds of implications of customization for after sales were not investigated), and the difficulty for after sales employees to distinguish the effects of customization from the basic nature of the spare part business (a high number of low volume items).

The limitation of this study is that it is founded on a single case. The replication of cases would provide a firmer base for generalizations. Furthermore, the effects of customization are presented as absolute figures. The lack of some kind of a reference value (for example, customization compared to standardization case by case) can be seen as a limitation. However, the detailed analyses made in the study

---

[34] A phenomenon that has not been the focus of this study should be kept in mind: assessment of customizations reveals that the selling volume of the customized items is critical in terms of costs. A low selling volume means more clarification work when the offer or order is handled. In addition, the design costs of customization cannot be covered in the case of low selling volumes of customized items.

enable generalization of the findings in environments that meet the conditions presented in this case.

Further research should be conducted to analyze the nature of the spare part business. It would provide a possibility to clearly dissociate the effects of customization from the effects of other strategies. On the other hand, an analysis of one or two case products in other companies would produce interesting reference to this study.

References

Åhlström, P. and R. Westbrook (1999): Implications of mass customization for operations management.: International Journal of Operations & Production Management **19**(3).

Coates, J. F. and M. F. Wolff (1995): Customization promises sharp competitive edge. Research Technology Management **38**(6): 6-7.

Computing Canada (1998): Virtual customer gaining power. Computing Canada. **24**: 6.

Darlington, J. (1999): Lean Thinking and Mass Customisation. Magazine for Chartered Management Accountants **77**(10): 18-21.

Deierlein, B. (1998): Good inventory management saves money. Fleet Equipment **24**(3): 42-46.

Evarts, E. C. (1999): More power to get what you want. Christian Science Monitor. **91**: 16.

Fitzgerald, B. (1995): Mass Customization - at a Profit. World Class Design to Manufacture. **2**: 43-46.

Fortuin, L. and H. Martin (1999): Control of service parts. International Journal of Operations & Production Management **19**(9): 950-971.

Gilmore, J. H. (1993): Reengineering for mass customization. Journal of Cost Management **7**(3): 22-29.

Gilmore, J. H. and B. J. Pine (1997): The four faces of mass customization. Harvard Business Review **75**(1): 91-101.

Jiang, P. (2000): Segment-based mass customization: an new conceptual marketing framenwork. Internet Research: Electronic Networking Applications and Policy **10**(3): 215-226.

Lampel, J. and H. Mintzberg (1996): Customizing Customization. Sloan Management Review **38**(1): 21-30.

Lapierre, J. (2000): Customer-perceived value in industrial contexts. Journal of Business & Industrial Marketing **15**(2/3): 122-140.

Lele, M. M. (1997): After-sales service - necessary evil or strategic opportunity? Managing Service Quality **7**(3): 141-145.

Lipe, J. B. (1995): Practice 21st century marketing. Corporate Report - Minnesota **26**(11): 22-24.

Mughal, H. and R. Osborne (1995): Designing for profit. World Class Design for Manufacture **2**(5): 16-26.

Pfohl, H.-C. and B. Ester (1999): Benchmarking for spare parts logistics. Benchmarking: An International Journal **6**(1): 22-45.

Quinn, J. B. and P. C. Paquette (1990): Technology in services: Creating organizational revolutions. Sloan Management Review **31**(12): 67-78.

Radder, L. and L. Louw (1999): Mass customixation and mass production. The TQM Magazine **11**(1): 35-40.

Sandvig, J. C. and J. J. Allaire (1998): Vitalizing a service parts inventory. Production and Inventory Management Journal **39**(1): 67-71.

Simon, H. and R. J. Dolan (1998): Price customization. Marketing Management **7**(3): 11-17.

Spring, M. and J. F. Dalrymple (2000): Product customisation and manufacturing strategy. International Journal of Operations & Production Management **20**(4): 441-467.

Sriram, V. and H. J. Sapienza (1991): An empirical investigation of the role of marketing for small exporters. Journal of Small Business Management **29**(4): 33-43.

Suomala, P., M. Sievänen, et al. (2000): The effects of customization on spare part business: a case study in the metal industry. International Journal of Production Economics Accepted: 10.

Transportation & Distribution (1999): SCM swoops into aerospace industry. Transportation & Distribution. **40**: 14-16.

Ulrich, K. T. and S. D. Eppinger (1995): Product design and development. New York (NY), McGraw-Hill.

Yin, R. K. (1994): Case study research: design and methods. Newbury Park, CA, Sage Publications.

Zairi, M. (2000): Managing customer satisfaction: a best practice perspective. The TQM Magazine **12**(6): 389-394.

The Authors

Aldous, K. J.
Industrial Research Limited,
PO Box 20-028, Christchurch, New Zealand.

Caddy, I.
School of Management, University of Western Sydney, Parramatta Campus
Locked Bag 1797 PENRITH SOUTH DC NSW 1797 Australia
Tel: 61-2-9685-9082; Fax: +61-2-9852-5647
E-Mail: i.caddy@uws.edu.au

Callan, J.
School of Marketing and Tourism, Central Queensland University, Rockhampton
Campus
Bruce Highway ROCKHAMPTON QLD 4702 Australia
Tel.: 61 7 4923 2550; Fax: +61 7 4923 2510

van Heck, E. & Vervest, P. H. M.
Dept. of Decision and Information Sciences, Rotterdam School of Management
Erasmus University Rotterdam, PO Box 1738, 3000 DR Rotterdam, the
Netherlands
T: +31.10.408.2032, F: +31.10.408.9010.
E-Mail: {m.wolters, e.heck, p.vervest}@fbk.eur.nl

Helou, M.
School of Marketing and International Business, University of Western Sydney,
Parramatta Campus
Locked Bag 1797 PENRITH SOUTH DC NSW 1797 Australia
Tel: 61-2-9685-9082; Fax: +61-2-9852-5647
E-Mail: m.helou@uws.edu.au

Hvam, L.
Associate Professor, Department of Manufacturing Engineering
Technical University of Denmark, Building 423, DK2800 Lyngby, Denmark
www.produktmodeller.dk

Knolmayer, G. F.
Institute of Information Systems, University of Bern
Engehaldenstrasse 8, CH 3012 Bern, Switzerland
E-Mail: knolmayer@ie.iwi.unibe.ch

Link, H.
VIAG Interkom GmbH&Co.
80260 München, Germany
E-Mail: link@genion.de
http://www.genion.de

MacCarthy, B. L., Brabazon, P. G. & Bramham, J.
Mass Customization Research Centre (MCRC), School of Mechanical, Materials
Manufacturing Engineering and Management, University of Nottingham,
Nottingham, UK

Nicholls, H. R.
Alchemy Group Limited
PO Box 2386, Christchurch, New Zealand

Österholm, J. & Tuokko, R.
Tampere University of Technology, Institute of Production Engineering
P.O. Box 589, FIN-33101 Tampere, Finland

Piller, F. T.
Institute for General and Industrial Management, Research Group Mass
Customization, Technische Universität München (TUM), Germany
E-Mail: piller@ws.tum.de

Rautenstrauch, C., Tangermann, H. & Turowski, K.
University of Magdeburg, Institute of Technical and Business Information
Systems
P.O. Box 4120, D-39016 Magdeburg, Germany
Phone: (+49) 391 67 1-83 86, FAX: -12 16
E-Mail: rauten@iti.cs.uni-magdeburg.de, Holger.Tangermann@gmx.de,
turowski@iti.cs.uni-magdeburg.de

Riis, J., Malis, M. & Hansen, B.
Ph.D. students, Department of Manufacturing Engineering
Technical University of Denmark, Building 423, DK2800 Lyngby, Denmark

Schackmann, J.
University of Augsburg, Business School, Department of Information Systems
Universitätsstrasse 16, 86135 Augsburg, Germany
E-Mail: juergen.schackmann@wiso.uni-augsburg.de
http://www.uni-augsburg.de/bwl/bwl_wi

Schenk, M. & Seelmann-Eggebert, R.
Fraunhofer Institute Factory Operation and Automation
Sandtorstrasse 22, 39106 Magdeburg, Germany
E-Mail: Seelmann@iff.fhg.de
www.mCustomization.de

Smirnov, A., Pashkin, M., Chilov, N. & Levashova, T.
St.Petersburg Institute for Informatics and Automation of the Russian Academy of
Sciences
39, 14th Line, St.Petersburg, 199178, Russia
Tel.:+7(812) 328-8071, Fax: +7(812) 328-0685
E-Mail: smir@mail.iias.spb.su

Suomala, P., Sievänen, M. & Paranko, J.
Tampere University of Technology, Industrial Management
P.O. Box 541, FIN-33101 Tampere, Finland
E-Mail: petri.suomala@tut.fi

Uuttu, O.
Modultek Oy
Lars Sonckin Kaari 14, FIN-02600 Espoo, Finland

Winter, R.
Institute of Information Management, University of St. Gallen
Mueller-Friedberg-Strasse 8, CH-9000 St. Gallen, Switzerland
Phone: +41 71 224 2935   Fax: +41 71 224 2189
E-Mail: Robert.Winter@unisg.ch

Wolters, M. J. J. (Corresponding author)

Printing: Strauss GmbH, Mörlenbach
Binding: Schäffer, Grünstadt